All Flesh
Is Grass

All Flesh Is Grass

The Pleasures and Promises of Pasture Farming

GENE LOGSDON

SWALLOW PRESS / OHIO UNIVERSITY PRESS

ATHENS

Swallow Press / Ohio University Press, Athens, Ohio 45701

www.ohio.edu/oupress

© 2004 by Gene Logsdon

Printed in the United States of America

All rights reserved

Swallow Press / Ohio University Press books are printed on acid-free paper ⊗ ™

12 11 10 09 08 07 06 05 04   5 4 3 2 1

Library of Congress Cataloging-in-Publication Data

Logsdon, Gene.

 All flesh is grass : the pleasures and promises of pasture farming /
Gene Logsdon.

     p. cm.

 Includes bibliographical references and index.

 ISBN 0-8040-1068-4 (cloth : alk. paper) — ISBN 0-8040-1069-2 (pbk. : alk.
paper)

 1. Pastures. 2. Forage plants. 3. Grazing. I. Title.

SB199.L64 2004

633.2'02—dc22   2004011948

To Richard Gilbert

Contents

Preface

To try in one book to appeal to the commercial farmer, the home food producer, and all consumers who seriously care about their food is not easy. But food production is in such a crucial state of bewilderment and ambiguity that I willingly take the risk involved. Commercial grain and livestock farmers languish in self doubt and discouragement because they must depend on government subsidies to make a decent profit. And the more they are driven by economics to seek profitability in quantity production, the louder the outcry from consumers worried about food quality. Harmful microorganisms like salmonella and listeria in conventionally produced animal foods have become serious, even deadly, dangers. Though highly exaggerated, the threat of mad cow disease (bovine spongiform encephalopathy) is scaring consumers away from conventional sources of beef.

Grass farming, or what I call pasture farming—what this book is about—is a solution to the dilemma. Unfortunately, most commercial farmers are as yet unable or unwilling to take the financial risk required to switch from factory farming to pastoral farming and not enough consumers realize what all is at stake. But a tide of information about this new kind of farming and the quality food it produces is rising. This book adds to it.

There is a compelling reason behind my writing. We live now in times as precarious as human history has ever recorded, at least mentally, which is the worst kind of precariousness. Beset by violence, economic instability, and ecological deterioration over which we seem to have no control, we feel helpless. The human race has become afraid of itself. In this quandary, people are trying to take their lives back into their own hands. Instead of mutely bowing before the mass psychology of violence and fear that infects society, individuals are taking steps to at least secure their homes—their food supply, their fuel supply, and their shelter. They are establishing homeland security not as political farce but in the vital, original meaning of the term. Most of them are moving out of cities, which seem more vulnerable than the country-side, but quite a few are also making their stand right within the urban world. It is for all these people, whatever their occupations, that I write this book.

In doing so, I must caution the reader not to expect much agreement among pasture farmers over the finer points of what they are doing. Their science is only in the developmental stage. And their art, like all art, is individual and particular. No artist worthy of the name ever completely agrees with anyone.

Giving Thanks

Richard Gilbert, to whom this book is dedicated, has been of enormous help, by supplying me with information he has learned the hard way in pasture farming and by being a longtime source of inspiration and support on all counts. Nathan and Kristine Weaver, Tim and Katie Kline, and David and Emily Hershberger, young dairy farmers on the frontier of pasture farming, have given me wise and inspired example and information, not to mention David Kline, Tim's father and one of the most influential persons in my life. It would be unfair not to also thank David's wife, Elsie, and daughter, Ann, who tell him what to do. I owe many thanks to my son, Jerry, and his brother-in-law David Grafmiller, for both their brains and their muscles as together we try to understand how sheep decide to do what sheep do. And I am grateful for my grandsons, Evan and Alex, for being able to outrun sheep now that I can't. Carol, my wife (who tells me what to do) and our sheep's midwife, is indispensable to our pasture farming. I am not only grateful for but also wondrously amazed at Bob Evans as he continues to challenge me to work as hard as he does at selling year-round grazing to farmers. Pasture farmers Bruce and Lisa Rickard and Mark and Debbie Apple, both by example and by direction, have helped in the writing of this book. Chelsea Monnin has been of more help than she realizes. Maury Telleen, Wendell Berry, and Wes Jackson have been there beside me at all times, enduring my bitching and moaning with patience and giving me reasons to be hopeful when I couldn't see anything to be hopeful about. I am especially indebted to Jim Gerrish, a very knowledgeable scientist and a proponent of pasture farming, for his suggestions and editing. I can always depend on neighboring farmers Dave Frey and Steve Gamby to set me straight when I go too far off into idealistic fancy. My sister Marilyn Barnes and her husband, Dennis, and my other sister Ann Billock and her husband, Brad, have all provided practical information about sheep and horses on pasture. And Marty Bender and Ed Vollborn have taken time to help with scientific oversight. Finally, I must always thank Steve Zender, Andy Reinhart, and Jan Dawson for their unflagging encouragement.

I have been blessed with the kindness and continuing support of Ohio University Press both from its director, David Sanders, and its senior editor, Gillian Berchowitz. Bless them both and all the people at the Press who work hard producing and selling my books.

Acknowledgments

Some of the material in this book has appeared previously in somewhat similar form in other books and periodicals.

The first chapter began as a speech delivered at Georgetown College, Kentucky, in 2002 and was then developed into an essay for *The Essential Agrarian Reader: The Future of Culture, Community, and the Land,* edited by Norman Wirzba (University of Kentucky Press, 2003). This essay I subsequently used as background material for another essay, "Farm History Is in Reverse," in *The History Channel Magazine* (September/October 2003).

Some of the profiles of farmers in the third chapter are revisions of profiles that appeared in the magazines *Farming* (Mt. Hope, Ohio) in 2002 and 2003 and *In Business* (Emmaus, Pennsylvania) in 2002.

Instructional material about grazing grains in the seventeenth chapter appeared also in an essay, "Marrying Grain and Pasture," in *The Land Report* (Spring 2002).

Faithful readers will recognize here and there a few humorous anecdotes that appeared originally in *Farming,* cited above, and in my columns in *The Draft Horse Journal* (Waverly, Iowa) and in my weekly newspaper column in *The Progressor Times* (Carey, Ohio) over the years. (There are plenty of new anecdotes, too, so don't despair, or do despair, depending on whether you think they are funny or not.)

In fact, this book is the result of all my writings on agriculture up to this point. I can go back twenty years and find the seeds of my ideas about pasture farming sprouting already. I thank all the editors and publishers who have so generously allowed me to use their publications as growing media for those ideas—even paid me! Since I may never get another chance, I would like especially to mention Joe Dan Boyd; Steve Zender; Jim Schley; Gillian Berchowitz; David and Carol Wiesenberg; Jerry and Ina Goldstein and their editor daughters, Nora and Rill; Maury and Jeannine Telleen and their editor son, Lynn; David and Elsie Kline and their editor daughter, Ann;

Jerry and Jill Carlson; Norman Wirzba; John Gallman; John Baskin; Jean Kelly; Barbara Ras; Craig Cramer; Tom Gettings; George De-Vault; Dick Braun; Scott Bontz; Terry Monahan; and, last but not least, Nicholas Roling, now gone, who, as moderator of my high school magazine, encouraged me so effectively to become a writer that, against all sanity, I actually did it. Alors.

1

Pasture Farming, the Newest and Oldest Agriculture

It may sound like agricultural heresy, but there are established, proven ways to produce most of the food that the world needs without tearing up millions of acres of the soil surface every year. We do not have to plant so much land to annual grain crops to eat well. In fact, the continued reliance on annual grains as the major source of feed for farm animals is proving to be the way not to eat well. We can have all the meat, milk, eggs, animal fabrics, and horsepower needed to feed the world without cultivating all those nice seedbeds that for centuries have been considered the beginning of the agricultural process by which human civilization survives. With modern grazing methods, farm animals can get their food mostly from grazing forages, thus avoiding the current crippling expense of annual cultivation and grain harvest.

I can make such an audacious statement not because I read it in agricultural journals, or because some learned scientist or earnest environmentalist told me so, or even because I have seen it demonstrated by other farmers (all of which reasons would apply), but because *I have been doing it, and by all that is holy, it works.* Moreover, it works best on the best land in America, the northern two-thirds of the nation from the east coast to the Great Plains, that area which we have always thought favored the production of annual grains. I have run out of patience with agribusiness grain monopolists who are in denial about the practicality of pasture farming over monoculture grain. They remind me of those clerics who refused to look into Galileo's telescope for fear of learning that they were wrong. We can junk all that expensive and destructive cultivation machinery, all those huge dinosaur grain harvesters, and the whole horribly expensive superstructure of grain storage and transportation. The days when that made sense are passing. That system is no longer profitable, as the billions of dollars of farm subsidies prove. History shows over and over that when a new way proves to be more economically profitable than the present way, it will be adopted.

I don't know why today's annual grain monopolists are surprised to see their domination of agriculture challenged by pasture farming. Annual tillage has been only a temporary blip on the screen of history. Even now, more of earth's space is used to produce food from pastures, tree groves, bodies of water, greenhouses, and mulched garden beds than from annually tilled fields. Most of the earth's surface is not amenable to annual tillage. Only in the last 150 years have landscapes of vast cultivated acreages become commonplace. Pastoral farming still reigns supreme where the land is too hilly, too dry, too wet, or too rocky for till farming. And now, pastoral farming is increasing on good arable land. Many graziers, particularly in New Zealand, Argentina, and the United States, could grow annual grains on their land successfully, but they have learned that by utilizing advances in agronomy, pasture farming reduces the cost of production dramatically and so nets more money per acre even if gross income is less.

I don't mean that we should quit growing grains altogether. But if the mix of 90 percent annual grain and 10 percent pasture that now reigns on our better farms was turned around to 80 percent pasture and 20 percent grain, farming could again be profitable and ecologically sane.

There are many types of pastoral farming, from ranching in Argentina, Australia, and our own West to the seminomadic herders of eastern Europe and the highlands of Asia. But today's grass farming, as it is commonly called, or year-round grazing, or managed rotational grazing, or, as I prefer to call it, pasture farming, refers to a kind of husbandry more fixed in place on relatively small acreages. The idea is to raise animals for food or fiber, or draft power or recreation, directly from pastures instead of penning animals up in confinement facilities and feeding them harvested grain. Pasture farming can also produce grains, fruits, herbs, mushrooms, and salad greens for direct human consumption as a sideline.

This kind of husbandry is being adopted by several different segments of rural society: seasoned commercial farmers who have concluded that growing annual grains or operating animal factories doesn't really pay; new farmers who see a chance to make money on a relatively small investment; part-time farmers who also have other jobs or careers; and gardeners who want to expand their food productivity beyond fruits and vegetables to meat, dairy products, and eggs. All of them recognize that pasture farming requires much less work and expense than growing annual grains and feeding them to animals in confinement. Furthermore, they understand that because of its ecological and environmental advantages, a pasture farm is also a joy to operate.

Pasture farming is the perfect entrée into husbandry for new farmers and gardeners who wish to establish themselves on small farms as a lifestyle move and want to make only part of their living on the land, not try to farm as a principal or sole source of income. (Very few farmers make their entire living from the land anymore.) Learning pasture farming requires no esoteric science. Anyone who knows how to establish and maintain a lawn has already learned the basics of how to establish and maintain a pasture. If a suburbanite divided the lawn into plots and stocked it with chickens or rabbits inside a predator-proof fence, he would, just that fast, become a pasture farmer.

Lest anyone think I am trivializing agriculture or being overly romantic by suggesting that nonfarmers, beginning farmers, and part-time farmers have a role to play in the future of food production, consider what is happening today. Mainstream agriculture languishes, kept going by billions of dollars in direct and indirect subsidies. A neighbor and I joke that there will be only two farmers left eventually: one on either side of the Mississippi. One will be Tyson Foods, and the other, Cargill, and eventually one will have to buy the other out because neither will be quite big enough yet to make a profit. We laugh, but not much.

Many economists don't laugh at all. They oppose subsidies to farmers and support a "free" trade policy that would move commercial agriculture out of the United States to Third World countries where, so they believe, food can be produced and processed cheaper. Responding to this kind of policy, John Berry, a farmer, former Kentucky state senator, and member of the Governor's Commission on Family Farms in Kentucky, writes in a letter: "there is a school of thought in this country that the battle is not between large and small farmers, or about whether farmers will remain independent or work for agribusiness corporations. It is about whether we will have any farmers at all."

If this kind of suicidal madness becomes government policy, anybody who owns or can rent even a couple of acres of land could become important in maintaining our food independence. I aim to make sure they have the knowledge they need. Already, newspapers and magazines are full of talk about how Brazil's twenty- to fifty-thousand-acre grain farms are going to run American grain producers out of business. Some American grain-producing industrialists are in fact moving their operations to Brazil. David Zartman, a leading scientific voice for pasture farming at Ohio State University, told me on the phone recently that he believes big grain farming will move to Brazil and other third world countries *in just one more generation!* He even said that I could quote him making controversial statements like that. He is not in favor of the United States losing its farms, which is why he espouses pasture farms and why I do, too. It is a way this country can compete if we are going to have a so-called global economy.

How Annual Grains Came to Dominate the Farm Economy

Civilizations have usually moved from pastoral agriculture to till agriculture as population increased. This transition has been so often observed that historians accepted till agriculture as the unavoidable result of population growth. That is, I think, a fundamental error. After long and arduous study of agricultural history, I am convinced that till agriculture has come to dominate food production in regions of fertile soil and plentiful supplies of water because of peculiar ways in which economics, climate, population, and culture have intersected, not because till agriculture possesses any inherent production advantages.

Not always have increasing population pressures led to more till farming. When the Enclosure Act in England allowed the rich to appropriate the public lands, the main idea was to graze sheep in the enclosures because there was a great demand for English wool in Europe.

For over a century, as population *increased*, there was much more land turned into pastures in England than into tilled fields because of the higher profits of wool over grain. (Interestingly, the pasture farming era in England ushered in the high tide of the British Empire.) In other words, if you want to discover the true cause of a trend in farming or food production, follow the money trail, not nature's trail. For two centuries annual till farming has been in the ascendancy because it suited the general money economy better. Now new economic factors are coming into play and they do not play well with giant till farms.

The human fixation with tilling the soil began as a convenient way to survive comfortably. Who wants to chase cows and pigs through the woods while being chased by wild animals and even wilder human marauders? The moats around castles were not for swimming. Towns were not walled because the walls looked pretty. Gardening and barn husbandry were started inside the fortifications not because of a lack of food beyond the walls but because producing food inside was safer and more reliable. The greater the danger out yonder, the more food production focused on tillage as a way to increase garden production in the small spaces available inside the walls. As the dangers decreased, till gardening moved outside the walls, where for no good reason except that it was familiar, it continued to be the agricultural choice.

Jane Jacobs, in her intriguing book *The Economy of Cities*, offers ample evidence that agriculture actually began in cities and then moved out into the countryside. The cultivated fields kept getting larger because of advances in tillage tools, not because farmers could not devise ways to get just as much food from pastured fields. Tilling seemed to work fine. If it ain't broke, don't fix it. Finally, constant tillage became not just the more acceptable farming method, but a cultural habit.

There is, in addition, a psychological reason why till farming gained ground. We humans understand how helpless we are in the face of nature's overwhelming power or even in the face of our own cruel and idiotic behavior toward one another. This realization, however unconscious, drives us to try to control everything around us. Our insatiable desire for money is a desire for control and security. The wisdom to accept limits to control is not part of our natures— witness the United States now as it wrestles with the heady temptation to rule the world. Till farming is control farming. Control farming is dreadfully expensive, economically and ecologically. Pasture farming, on the other hand, requires a humble dependence on forces beyond tillage machines and a recognition that humans are not really in control, no matter what. Pasture farming recognizes that our survival depends upon our ability to stand by patiently and work in partnership with nature, not in domination over nature.

The most insane example of till farming as cultural habit occurred (and is still occurring) in our own Great Plains, where vast acreages of prairie grasses were plowed up and turned into a dust bowl. Even today the dry plains country remains profitable for annually cultivated crops only because of irrigation, using, in many cases, fossil supplies of water. Had that land been left in grass, and had forage crops been improved to carry more animals per acre, this whole area would be thriving today instead of surviving as a huge no-man's-land of subsidized corn and soybeans dotted with decaying towns and farmsteads. Humans know that; they just can't yet acquiesce to the truth of it. Northern China is now making the mistake all over again, creating unimaginably vast areas of cultivated soil that turn into dust bowls when the wind blows. It will happen eventually in Brazil, now viewed by agribusiness as Tractor Heaven.

Not that there hasn't been evidence all along that pasture farming might be just as promising as till farming. In 1775 James Anderson, an English agronomist, described in his journals the same practice of strip-grazing that has been resurrected by today's graziers. Wrote Anderson: "A farmer who has any extent of pasture ground should have it divided into fifteen or twenty divisions . . . and the beasts be given a fresh park each morning, so that the same delicious repast might be repeated" (*Essays Related to Agriculture and Rural Affairs*, 2d ed. [Edinburgh, 1777]). But by then Jethro Tull's cultivators and drills were all the rage, even though Tull's theories about why plants should grow better under cultivation were ludicrously wrong. As he described in his *Horse Hoeing Husbandry*, published in 1733, he believed that plant roots had little mouths on them that swallowed microscopic particles of soil. The more the soil was broken up and pulverized, the more tiny particles could be pressed into the roots.

André Voisin, in his classic pasture book, *Grass Productivity*, stated that rotational grazing did not carry the day in the 1700s because Anderson and the graziers who followed did not understand that the timing of rotations was even more important than the number of animals in the rotation. As much as I admire Voisin's work (everyone who cares about health and food should read his *Soil, Grass, and Cancer*), I believe he was judging earlier graziers too harshly. Any husbandman with half a brain is going to realize after a few years of experience that the timing of the rotational moves must vary depending on how the pasture is growing and its need to regrow before being grazed again. Even I figured that out on my own. Rather, I think grazing lost out for psychological reasons. The new cultivating gadgets got all the attention because they were something that could be manufactured and *sold*. Money moves mankind. Shifting animals from pasture to pasture didn't stand a chance of being championed in commerce because there was little manufacturing wealth to be derived from it. Besides, very soon farmers could *ride* on their cultivators and till to their heart's content. And so they did, unto this very day, because humans love to move around while remaining motionless on their butts.

The trend in husbandry in England in the 1700s and 1800s followed and increased the focus on till agriculture. The Enclosure Acts forced hundreds of thousands of subsistent farmers off the land and into cities. They became the new consumer market for the livestock products of wealthy landowners who co-opted the enclosed lands. Breeds of livestock at that time were wondrously diverse, but half wild, often wholly wild. Improving that stock in size and rate of gain for meat to sell to the dispossessed workers now earning money in the new factories promised good profits. The average weight of beeves at the great Smithfield Market in London went from 380 pounds to 800 pounds between 1710 and 1795. (I rely on a most revelatory book, *Farm Animal Portraits*, by Elspeth Moncrieff with Stephen and Iona Joseph, for these details and others that follow.) Other farm animal species increased in size and weight accordingly. But the practical side of this improvement of breeds soon gave way, as all human endeavor inevitably seems to do, to impractical extremes fueled by the desire for short-term gain. The wealthy "gentlemen" landowners, like those in the United States in the latter half of the twentieth century, were soon caught up in the competition for prizes granted at fairs. The judges awarded prizes mainly on size, thus encouraging breeding of still larger animals. The competition increased enormously, not just because of the profits that accrued to the winners for their breeding stock, but from the all-too-human desire to beat rival breeders in the show ring. Because the participants in the competition were mostly wealthy, money was no consideration. Before long, all practicality as to breeding useful and efficient farm animals was abandoned in pursuit of prize-winning appearance.

Eventually, gentlemen farmers were raising animals of enormous, almost unbelievable, size: oxen weighing two tons, sheep of four hundred pounds or more, and hogs of over a thousand pounds. So ludicrous did the competition to win prizes become that hogs were bred and fed to such ballooning fatness that they could barely stand and when sleeping had to have their snouts propped up with wooden "pillows" to keep them from suffocating in their rolls of fat. (This practice is thoroughly documented. See the illustration *The Cup Pen of Pigs*, engraved by E. Hacker, 1874, in *Farm Animal Portraits*, page 242.) If a hog could walk two hundred feet at a time, said William Cobbett, the famous farm writer of that era, it was not yet fat. He meant that as a compliment to fat breeding, but admitted that the sow he raised to enormous size would not breed, a common fate of overfeeding even today, especially true of bulls and rams raised for the show ring rather than for practical husbandry.

Although there was at the same time practical improvement of breeds going on, it was the show ring and the prizes, and the subsequent fame and money, that drove "progress" in animal "science." Such animals could not graze efficiently, could barely walk, in fact, and so they had to be coddled in barns and fed exotic diets of grain and vegetables and supplements that would allow them to put on the enormous weights that breeding up the size of the animals made possible. *This practice reinforced annual tillage because the fattening feeds had to be raised by cultivation rather than by grazing.* The name of the game was fame and money for the few, not practical, low-cost farming for the many.

Needless to say, breeding and feeding for size led to the same problems that it does today. When the Cheviots were "improved" to larger size by crossbreeding, they lost their hardiness and in the severe winters of the 1870s whole flocks were wiped out. Similarly, today, we have developed hogs for factory production that can't survive outdoors in winter. When fine-wooled sheep were "improved," their wool became coarser. Shetland wool, renowned around the world in the seventeenth century for its softness, lost its market. Practical farmers eventually went back to the original breeds to some degree, but in the meantime the wonderful diversity of livestock, with each county having its own breeds particularly suited to local climate and food supply, was decimated by crossbreeding or abandonment.

By the middle of the 1800s, complaints about the fattiness of the meat from "improved" livestock were being voiced far and wide, and reaction set in. One writer observed in *British Husbandry* in 1834 that the meat was "too dear to buy and too fat to eat." Another, writing in 1805 and quoted in *Farm Animal Portraits,* describes a leg of "prize" mutton thusly: "the fat which dripped in cooking was measured and it amounted to between 2 and 3 quarts, besides which the serving dish was a bog of loose, oily fat, huge deep flakes of it remained to garnish that which we called, by courtesy, lean, being itself also thoroughly interleaved and impregnated [with fat] . . . . Little of it was eaten at our table."

Told that his meat was not fit to eat, Robert Blackwell, the famous breeder of "improved" fat livestock at that time, replied tartly that he did not produce meat for the tables of gentlemen but for laborers. Presumably, laborers could not afford good meat and didn't know the difference anyway. One can't help but note that Blackwell's attitude is reflected today by the factory meat industry, which in spite of universal protest from nutritionists and health-minded consumers, continues to provide fast-food restaurants with mountains of fatty, corn-fed meat because "that's what 'thuhmarekinpeople' want."

Two paintings in *Farm Animal Portraits* bear particular meaning for today. One, executed in 1810, shows a grossly fat bull, a grossly fat pig, and grossly fat sheep in front of which stands the grossly fat owner, his huge, protruding potbelly exactly matching the animals'. The other shows a fancy butcher shop of 1822. Amid the fat-laden cuts of hanging meat produced with the animal feeds of annual tillage is one that seems a little leaner. On it is a sign that one might very well find in an upscale butcher shop of today. The sign reads: "Grass fed."

We have tilled the soil so long that we can't imagine not doing it. The plow and the amber waves of annually planted grain from sea to shining sea are our cultural icons. Wes Jackson, a plant geneticist who at his Land Institute in Kansas is demonstrating the possibility of a practical, permanent, till-less agriculture, wrote the following in his 1987 book, *Altars of Unhewn Stone:* "From the moment humans first touched plow to soil, exchanging hunting and gathering for domestic agriculture, we committed ourselves to ultimate decline. It is a tragedy in Alfred North Whitehead's sense of tragedy, the remorselessly inevitable working of things. Given the current human population now dependent on till agriculture, we will need to continue to till the earth even though such activity has historically and prehistorically undercut the very basis of our existence."

Why the Seeming Inevitability of
Till Farming Ain't So

But the "inevitable working of things" can work the other way, as Dr. Jackson and now thousands of farmers are betting their careers on. The way that money and fossil energy are being spent to grow industrial grains that are then fed to confined animals is perilously extravagant compared to pasture farming. It takes only a cursory comparison to prove the point. In till farming today, farmers go to the field in spring with very expensive machines and weed killers to prepare the soil and plant the crops. Ironically, the much-vaunted "no-till" farming of recent years that uses herbicides for weed control rather than soil cultivation at least proves the contention that crops can be grown without tearing up the soil surface.

But there are problems with no-till in annual tillage (it works great for planting new grasses and legumes in an undisturbed pasture sod), and many farmers have gone back to more tillage, not less, starting erosive soil preparation in the preceding fall. Farmers where I live do it, particularly for corn, because, following the law of averages, there is only a relatively short planting time in spring to gain the good yields necessary for a profitable crop. Rains make the chances of getting the crops in during that optimum time a challenge, and a challenge that looms larger as the farms get larger. If the soil has been fluffed up (laid bare) by cultivation in the preceding fall, it will dry out two or three days faster in spring, making possible that many more hours of optimum planting time. Farmers are willing to gamble erosion, time, and fuel in the fall on the chance of gaining those few precious days. Enormously large tractors, cultivating equipment, and planters are therefore necessary to plant huge acreages fast. Moreover, if weather tends to be wet in April and May, the fitting and planting will go forward of necessity when the soil is not dry enough under the surface, causing deep compaction.

But whether by "no-till" or "overkill" methods, the grain crops are finally in the ground and maybe start to grow. (The word *maybe* is necessary here. In 2002, large-scale farmers put corn in the ground in April here in Ohio to take advantage of an early break in rainy weather. But temperatures continued cold and not enough of the kernels germinated. Replanting was necessary. By then the optimum planting time had passed and dry weather ensured a bad crop.) Then the farmer must worry over the possibilities of hail, flooding, or drought.

More spraying of pesticides may become necessary, or deemed to be necessary, because of weeds immune to the first spray or unforeseen bug and fungal infestations. Banks that have loaned the farmer money may demand spraying to "protect its investment." In any event, most industrial grain farmers now carry crop insurance, another expense (around fifteen dollars an acre), because as the size of farms grow, the greater the vulnerability to weather-related problems.

Finally comes harvest. Now the farmer goes to the fields with enormous harvesters costing $200,000 or more and expensive semi trucks to transport the grain to market. If the autumn is wet, harvesting will be delayed or finally done anyway on wet soil that again will compact badly under the weight of heavy equipment. I've ridden in giant harvesters right through half a foot of water as the cutter bar skimmed off grain heads above. Meanwhile, the big semis, thundering up and down country roads never made for them, tear up the roadbed.

The grain must almost always be artificially dried, using natural gas, stored, and kept free of pests in on-farm storage or at grain elevators, all at great expense. Then the grain must be shipped out again, by truck, or train, or barge, or a combination of all three, at more expense in fuel and road degradation, to incredibly expensive animal confinement facilities. I'm told that it is cheaper now for chicken factories on the East Coast to buy their grain shipped by boat from Brazil than by train, truck, and barge from our own midwestern grain farms.

The costs continue to skyrocket at the animal confinement facilities. Their mills to grind the grain into feed are enormously expensive. Handling the manure is enormously expensive. The employees, often immigrants, may not be well paid, but the labor bill is still considerable. The hormones and antibiotics necessary to keep the factory animals growing aren't free either and create health risks for both animals and humans. Workers wear masks in hog factories to fend off the dust that causes what is called "farmer's lung." Then the meat, milk, eggs, or other products must often be shipped back to where the grain came from.

Needless to say, the amount of energy involved in all this is horrendous. The total amount of fossil fuel burned to manufacture all the machines for industrial grain farms and animal factories is practically incalculable when one adds into the equation the energy used in building the trucks, trains, automated feeding, and manure handling equipment in the animal factories and the confinement buildings themselves.

Compare that to a pasture economy, where in place of all that madding crowd of toil, the grazier turns his animals out to graze. Assuming that the grasses and clovers and other grazing crops have been established, spring work amounts to moving the animals when necessary to new grazing and keeping the fences in repair. Rains do not hamper the operation because there is no soil cultivation. Likewise, there is little erosion. Hail will not harm the pasture the way it will a grain crop, and flooding only temporarily. Some graziers I know well do not have barns for their livestock at all. They say that the necessity of barn shelter is a myth. There is no zillion-dollar cost for machine harvesting. The animals do the harvesting, apply their manure for fertilizer, and eat most of the weeds. The cost of added fertilizer and pesticides is less than in till farming and continues to decrease the longer a pasture is managed properly. The only machinery necessary is for mowing and making hay. If the grazier becomes skillful, he will have to make hay only in early summer for feeding in winter. When the grazier gets *really* good, he can graze year around, even in the north, and hold enough hay in reserve for only the deepest snows and, in some soils, during spring thaw when the ground is too soft for trampling by animals. As a final gift from sane agriculture, pasture farming has a far longer season of production than annual grains. In 2002, the driest year on record here in our part of Ohio, the corn and soybeans (what there was of them) were finished by September while pastures were tuning up with fall rains to give us three more months of grazing.

It doesn't take a genius to figure out which farming method is more economical for the farmer. But there are other advantages that will make pasture farming look even better in the future.

1. Cash grain farming lives on cheap fuel. Fuel is not likely to remain cheap.

2. Machinery costs will continue to mount. A system of food production that is almost immune from what graziers call "heavy metal disease" will have a decided economic advantage.

3. Pasture farming can utilize soils profitably that are too poor or erosive for profitable annual grain farming. As competition for land increases (housing developers and road builders like deep, well-drained soils, too) this fact will be a critical advantage for pasture farming.

4. The latest research reports say that annual soil tillage contributes to the buildup of greenhouse gases, notably CO_2. If that is true, score another plus for pasture.

5. The corn and soybean monopoly looks to ethanol as its life-saver, but studies like those done at the Land Institute in Kansas suggests that when all is said and done, it takes about a gallon of ethanol to make a gallon of ethanol. Moreover, genetic engineers are saying that they will be able to produce ethanol more economically from improved perennial grasses than from corn. The only reason corn is being pushed for ethanol is that it is the crop most in surplus right now. As David Pimmental at Cornell points out, the more corn is used for fuel, the higher food prices will rise.

6. It is extremely doubtful that animal factories will be able to achieve a profit without continued cheap subsidized grain and subsidies for manure management. Attempting to handle very large quantities of manure (as much as a large city, in many instances) in what is essentially a flush toilet system just doesn't work well without a costly sewage treatment plant or a composting plant. Pollution calamities from liquid manure holding ponds and pits at animal factories continue to demonstrate that conclusion. When large confined animal operations are required to install pollution controls as the rest of industry and city sewage systems already must do, their costs will rise farther above pasture-raised animals. Pasture farming requires no manure management system. Even the practice of dragging a harrow over field droppings to spread them out is not necessary in most cases because it can spread the "zone of repugnance" in which the animals won't graze. Soil bacteria, earthworms, dung beetles, and other organisms will break down droppings better and faster.

7. Animal confinement facilities have been so plagued by disease pathogens that the government and the meat processors are moving ahead on irradiation of meat. Irradiation, says *Pro Farmer Connection* in its November 21, 2002, edition, destroys 95 percent of vitamin A in chicken *before* you cook it. Irradiation increases mutagens and carcinogens like formaldehyde and butane, says Roswell Park Cancer Institute's Dr. George Tritsch. (I'm still excerpting from the same source.) Irradiation kills beneficial organisms that check botulism. As *Pro Farmer Connection* points out, what this all means, among other things, is that pasture-fed meat will surely sell at a premium because knowledgeable consumers will refuse to buy irradiated products. That's why the industry is starting to use the term *cold pasteurization* instead of *irradiation*.

8. Other significant health benefits from eating animal products raised in a grazing regimen will mean premiums for the pasture farmer.

Alan Nation, in the September 2000 issue of his *Stockman Grass Farmer,* writes: "in twenty years, all of the grass dairies and grass finishing operations [will] be on Class I farmlands. The same quality soil that will grow 185 bushels of corn [per acre] is the same quality soil that will best grow the necessary . . . dairy and beef finishing forages. New Zealand's famous Waikato Dairy District and Argentina's beef finishing zone [both based on pasture-only farming] are those countries' best land, not their worst . . . [But] the naysayers ask, 'Why go to all that trouble when you can just feed grain for the extra energy?' There are a lot of arguments for pasture only, but today I will use the simplest one. Grain feeding is going to increasingly prevent you from taking advantage of premium-priced market options. Research has found that feeding as little as five pounds of grain per day cuts the CLA [conjugated linoleic acid, which holds great promise in fighting cancer, obesity, and diabetes] content of the milk in half! Also in research, grass-finished beeves had 250% more CLA in their intra-muscular fat than grain-finished beeves."

9. Meat and eggs from animals grazing pasture compared to animals eating heavy grain rations also have less cholesterol, less fat, more essential omega-3 fatty acids and several anticancer agents, and more essential vitamins. (Jo Robinson's *Why Grassfed Is Best,* among other studies, details and documents these claims.)

Change Comes Slowly When Big Money Feels Threatened

With all these reasons for a change in our agricultural focus, and with a significant number of farmers experimenting successfully with that change, why all the apathy from mainstream agribusiness? The first and most significant reason is that because of the way pollution regulations are being applied to confinement operations, or rather not applied, the only limit on the number of animals you can cram into a small space is the amount of land you have available for more buildings. A pasture farm can support only a certain number of animals per acre. This is really the whole story of why agribusiness thinks pasture farming can't be as profitable. Confinement and concentration make it appear that there is a possibility of open-ended quantity of production and therefore open-ended wealth. This kind of thinking avoids taking into account the fact that an acre can still produce only so much.

Second, even farmers who would like to turn to pasture farming often have so much money invested in machinery and buildings that they think they can't financially make the change.

Third, because pasture farming does not require many of the inputs of till farming, agribusiness sees it as a threat to profitability. The threat goes very deep. The farm economy rests upon an extravagantly expensive economic infrastructure. Thousands of jobs would be lost, agribusiness fears, if pasture farming suddenly became the dominant force in farming. But if our present farming can't compete with Brazil, as these same agribusiness people claim, the jobs will be lost anyway. At least pasture farming would generate some new jobs. Which would you rather do, work on an assembly line or move livestock from one field to another?

A common argument in favor of grain farming over grass farming is the ingrained belief (oh dear, pun) that more food for humans can be produced per acre with harvested grain fed to animals than with grazing. Is this claim true? It has certainly not been proven so. If agronomists had spent the same efforts on improving plants for grazing as they have for grain crops over the past two hundred years, I am quite sure it would not be true. But just for the fun of arguing the point, let us consider the best, levelest, deepest Illinois soil that can consistently produce two hundred bushels of corn per acre if you sock the fertilizer to it. That same acre managed the same way can produce ten tons of a mixture of alfalfa and orchard grass. Which crop produces the most animal food product for humans?

Marty Bender, a scientist with the Land Institute in Kansas, was kind enough to do some arithmetic for me. "I think I can make a very strong quantitative case for your pasture farming," he wrote back to me.

> Generally speaking, major grain crops are not superior to major forage crops in the amount of total plant growth. In particular, the two crops you are considering [corn versus alfalfa/orchard grass] produce about the same amount of total plant growth. But pasture grazing gets a tremendous advantage in feed yield because it consumes most of this plant growth, whereas corn harvest removes only about half of it in the form of grain. Now this does not mean that the pasture will yield twice the animal products as the corn crop because we all know that for a given weight of feed, ruminants and pigs

gain more on grain than on forage. But with this much advantage in feed yield, it is surely conceivable that pasture grazing could produce as least as much animal products as grain. In fact for your comparison, I looked up tables of feed composition and calculated that the pasture provides *more than twice the feed protein and the same amount of digestible feed energy as the corn grain* and that applies to both ruminants and pigs.

Also, Ed Ballard, a University of Illinois extension specialist, concludes from his research findings that profits from grass farming are higher than from cash grain on the best corn land in Illinois. It would appear that only government programs keep Illinois corn farming afloat.

I rest my case.

I would add that when Mr. Bender says animals gain more weight on grain than on grass on a given weight of grain, which is true, what that really means is that the animals gain *faster* on grain up to a point. This is an economic argument, not a nutritional advantage. The faster the animal grows, the faster the money can be turned over. Faster gain has nothing to do with nutritional gain. There is no law that says a hog must have a diet of corn to grow healthfully and produce good pork. Hogs go wild and stay very healthy without a kernel of domestic grain. I daresay that gourmet hams from hogs finished on a diet high in acorns, which lends a unique taste to the meat, taste better than "corn-fed." My father-in-law would not think of smoking hams from fast-food hogs. He wanted hams from a hog fattened over twice the time that market hogs are fattened.

But of course the debate over taste of grain-fed versus grass-fed is not solvable because taste is so subjective. Many supposedly scientific taste tests have been done with varying results. In recent years, more tests have concluded in favor of grass-fed than formerly because, I suppose, the meat used comes from pasture farms employing the latest management techniques, not from hillsides covered with poverty grass. The corn industry responds that when people can afford it, they buy choice or prime beef, fattened on corn. That is no longer always true. There is a sharp demand for grass-fed beef now.

I think this taste debate is ridiculous. Certainly, the best of both grass- and corn-fed are very good if you like meat. Method of cooking is at least as important as method of feeding. The only notable difference in taste quality that I discern in various meats has nothing to do with how the animal is fed. All meat from all animals tastes better to me if it is produced and processed at home or at small local slaughterhouses. Why home-grown and home-butchered meat invariably has a better taste I can't say for sure, but I believe it is because home butchers and local slaughterhouses age the meat a week to as much as three weeks, while in industrial plants the time is usually around thirty-six hours.

Preference depends mostly upon what one is used to eating. Some people swear that Jersey meat is the best; others eye the yellowish color of the fat and make up their minds before they taste it that they don't like it. I can taste no difference in the meat from different breeds if all were fed the same way, butchered at the same age, chilled and aged the same way, and cooked the same way. Once we had some Piedmontese beef, touted for its low cholesterol leanness. It was grown to regular market weight on only fair pasture and hardly any grain. It was tough. But marinated, it was fine. Certainly it had a good taste. I've a notion that the marinade may have had as much to do with that as the meat.

Wendell Berry writes a most telling observation about beef taste in his book *Home Economics,* concerning his 1982 visit to Ireland:

> A sign on the wall of Johnny Flood's butcher shop caught my eye: "Great grass makes great beef." This sign is distributed by the Meat Marketing Board, which has mastered the not overly difficult truth that in a grass-growing country [much of Ireland is not amenable to annual cultivation] the people should eat grass-fed beef. In the United States, also a country with extensive areas superbly suited to grazing, the era of cheap petroleum has produced a long-term glut of cheap corn, ruinous both to farmland and to farmers, and this has in turned produced, according to certified experts, an inflexible preference on the part of "the American housewife" for corn-fed beef. So much for the preference of "the American housewife," I thought, looking at the sign in the Oldcastle butcher shop, "who so uncannily prefers exactly what the American meat industry prefers that she should prefer."

Generally, the choicest, priciest corn-fed beef tastes a little too fat or tallowy to me, but it is still very good. However, I like baby beef better, a calf raised on plenty of mother's milk and choice grass and clover without any grain to about 650 pounds of weight and then butchered at home without going through the trauma of being trucked, pushed, and prodded into killing lines and slaughtered in the bawling pandemonium of packing houses, an experience that elevates adrenaline flow in the animal and, according to gourmet meat lovers, affects taste of the meat adversely. More people might agree with me, I think, but there is rarely any real baby beef in stores and so consumers don't know what it tastes like. It is as tender and juicy as any choice grade beef and has what I call a live taste that doesn't fill me up unpleasantly like highly marbled prime beef.

And here's a tip worth more than the price of this book if you raise your own meat. A grizzled old cowboy from Nebraska once told me that the very best steak comes from a four-year-old range cow, because, he said, it takes four or five years to put some real flavor in the meat. But isn't it as tough as shoe leather? I asked. Not if you eat a steak from the strip of loin along the backbone (where filet mignon comes from). The loin doesn't get unpleasantly tough even in an animal that old, he said. I checked that out with a professional butcher. He agreed. So now when I take to the slaughterhouse an old cow that has come to the end of her productive life, I don't have *all* the meat ground into hamburger like I used to do. (Old cow makes very tasty hamburger, by the way.) I have the loin stripped out first for filets and then have the rest ground into hamburger. If you do likewise, your butcher will be most impressed with your astuteness, even though it means he can't be tempted to abscond with a few filets for himself—perish the thought.

The general arguments I have given in support of pasture farming are sort of déjà vu to the student of American agricultural history. From about 1947 to 1957 there was a tremendous groundswell in favor of pasture farming, referred to then as grassland farming. Many books were written on the subject. Possibly the most popular was Channing Cope's 1949 book, *Front Porch Farmer,* in which the author tells how he learned to farm his seven hundred acres in Georgia mostly from his porch while his cattle did the work. Farm magazines were full of articles singing the praises of grass farming. I have found the ones in the *Farm Quarterly,* which itself has long since passed into history, most helpful in learning how to manage pastures today. The only magazine that has persisted in reporting on the progress of grass farming is the

Stockman Grass Farmer, and it has become, to no surprise of mine, one of the few farm magazines increasing in circulation. I highly recommend it. You will find documentation in it for most of the "outrageous" claims I make in this book. A brand-new magazine, *Farming,* also carries many articles on grass farming, all written by farmers who are practicing it.

But the question remains: if grass farming is such a great idea, why did it not become the dominant form of farming by now instead of declining at the end of the 1950s, not to march back into the limelight until the 1990s? My easy answer is the same one I give for any event in history that does not make sense to me: the general insanity of humanity. However, there were other reasons why it did not take hold then. The economy of the late forties and the fifties still seemed, on paper, to favor annual cultivation of grains. Grain farms were still comparatively small and did not buy as many supplies from agribusiness as now. Herbicides were just in their infancy. More of the work now hired out to agribusiness custom services was done by farmers who were accustomed to receiving minimal returns on their labor because the farm supplied some of their costs of living and much of their recreation. In general, they did not yet make use of so much high-priced, off-farm labor as they do now. Grain could be grown at less cost per acre in terms of machinery and fertilizer. Grain was still mostly air-dried, stored, and fed right on the farm, avoiding many of today's high costs of elevator and mill processing. Much of the meat, dairy products, and eggs was sold locally. Long-haul, refrigerated trucks and airborne transportation were only beginning. There was as yet no interstate highway system. (Heavens, there were no big-league baseball teams on the West Coast yet.) In those days, harvested grain could compete with pasture in meat and milk production.

And then the government made the Great Decision. Grain meant more jobs—at least off the farm. Grain was easier to transport, store, trade, and sell as a commodity far from the farm. Grain could and did become coin of the realm, a way not only to consolidate wealth by controlling food markets here but also to gain economic and political advantage by exporting to other countries. Because of these grain "advantages," the government subsidizes grain farming and until recently not pasture farming. Grain farming is the way to concentrate the control of agriculture into an oligarchy of the wealthy. Had it subsidized the growing of grasses and clovers as assiduously as it did and does annual grains, we might have avoided the heavy grain surpluses that continue to ruin the farm economy. We might have avoided the artificially low prices that in turn required even more subsidies. It is interesting that in New Zealand, where the science and practice of grass farming has advanced the most, government subsidies have been abandoned with no bad economic effects to the country and, in fact, with improvements to the farm economy. (See almost any issue of *Stockman Grass Farmer* for verification.)

But subsidies to keep food cheap and off-farm jobs plentiful seemed and seem like a better goal to American leaders. Economic policy forces farmers to consolidate into ever-larger units, driving millions of people out of farming. The result has been anything but cheap food if all costs are taken into account. And many of the jobs that this policy has fostered no longer translate into profitable work.

Perhaps it could not have been otherwise. Can humans voluntarily give up the kind of greed that leads not to great efficiency from consolidation but to gross inefficiency? Other than the few who live on the philosophical ramparts and in every age try to point out the dangers of the consolidation of power, who would have believed that encouraging smaller farms based on pasture farming instead of huge farms based on manipulated money interest rates would have in the end led to more jobs immune from layoffs, to truly cheaper food, and to a much healthier distributive economy? Now this new economy is coming the hard way, and not all the king's subsidies nor all the king's technologies can put the broken Humpty Dumpty of too much till farming back together again. Not in a free country, anyway.

2

How I Came to Pasture Farming

Our farm is tiny, just thirty-two acres of which only fifteen are devoted to pasture farming. The rest is in woodland, gardens, and fruit trees. My son and his brother-in-law are also beginning pasture farming on their little farms not far away. We all work together, and what I've learned so far comes from our collected experience (three heads are better than one, not to mention three wives' heads, which are even better) on a total of about thirty acres actually devoted to grass farming, with more acres to be gradually added. Thirty acres hardly make a commercial-sized pasture farm, but what we are doing could work on a farm of up to a couple hundred acres with only a modest increase in haying machinery, more fencing, and more water tanks. Most important, we do nothing that any family of modest income can't do. I speak to and for ordinary people doing what ordinary people can do with ordinary financial means. We have gotten to where we are today without high incomes, inherited money, grants, or pie-in-the-sky schemes for making a pile of money. If that were not so, I would feel, in writing this book, like an agricultural con man, of which there are already quite a few. I can't tell you how to make money. No one can. Making money is an art and skill in itself and sometimes an obsession. Money success belies any step-by-step procedure that can be laid out like a puzzle for a person to fit together. Two people can pursue what appears to be exactly the same course. One makes money; the other doesn't. One of America's great myths is that its universities can teach a person to become wealthy. It is all a scam. It would be about as easy to teach a hippopotamus to fly as to teach a farmer how to get rich, mostly because hippos aren't interested in flying and most farmers aren't interested in getting rich. After many years on the land, I know of only a few farmers who have made a lot of money from farming. For the few who did, it is more a matter of timely luck than method. If nature gives a farmer a break now and then and the farmer is wise, skillful, and very careful with money, he or she makes enough for a modestly comfortable living. What money is accumulated in addition comes from the rise in the value of farmland or from investments that prove in hindsight to be smart ones. Such money does not come directly from production farming. I speak to the people who understand that and are content to live and work with that reality.

I think of our endeavor two ways: first, as the kind of farm many part-timers with two to twenty acres can find practical and satisfying while providing themselves with a little or a lot of their own food, fuel, fiber, and recreation; second, as a large test plot for trying out new and old ideas that would apply to commercial pasture farms of any size. I think that what we are doing has possibly more informational value than a university trial, because ours, while not rigidly scientific, is a more holistic model embracing an entire farming system rather than one or two isolated practices. Our farming is not, in any event, a hobby farm, a term that makes me bristle when it is applied to our work. This is no hobby, except in the sense that I can follow my hobbies of gardening, nature-watching, and hunting Stone Age artifacts while I experiment with pasture farming.

There are two levels at which grass farming can be practiced. The first is professional, commercial grass farming, which, as I will point out probably more often than you care to read, operates in a marketplace that is biased in favor of professional, commercial grain farming. To compete in that market, especially in dairying, grass farmers must manage pasture rotation and strip-grazing in very sophisticated and exacting ways. This kind of grass farming is proving successful, but it is as demanding as any other kind of commercial enterprise.

The other kind of grass farming, the one we follow, is less intense, less demanding, and quite easy to accomplish because it does not rise or fall on market realities. Nor does getting into it demand much money. It not only avoids all the risk of present-day grain and animal factory farming, but also most of the hard challenges that commercial grass farming must meet and solve. For example, we don't have to worry much about livestock traffic over the pastures the way graziers of large herds must. At our level, we hope to make a little money, but our main goal is to market what we produce to ourselves and a few select customers. We figure that we "sell" at retail supermarket prices and so always can show at least a little profit. For example, the three hundred pounds of beef we raise to eat, we value at store prices of about three dollars a pound, which is considerably more than we'd get for our baby beef if we shipped it to the stockyards. Our kind of grass farming, or grass gardening, as I am tempted to call it sometimes, compares to commercial grass farming in the same way that gardening compares to commercial truck farming. But unlike the transition from gardening to commercial horticultural production, the garden farmer can, if he wants to, move from a very small grass farm operation to a larger commercial one without any big expense in machinery or other inputs. Grass farming is brain farming, not factory farming.

I also have another goal. Not to sound overly dramatic, but I have a vision of an agriculture that could produce good food and plenty of it from biological power *alone*. I am not going to reach that goal in my lifetime, but having it as a possibility inspires me to find ways to use less and less fuel and heavy machinery. I think there is a distinct possibility that lower- and middle-income people will have to produce food in the future mostly on muscle and brain power, because nonbiological power will be too expensive.

My experiences as a grazier are ongoing works in progress. We have not yet developed anything close to the goals that we know are possible, nor have we achieved the expertise of the top commercial graziers. I am not even sure that we are totally on the right road yet; there are many ways to cook this pasture farming goose. We do not follow the highly sophisticated and skilled grazing methods of the real pros who move dense populations of animals over small paddocks very frequently, even every day, to get gains in pounds of meat or milk that will allow them to compete in the grain-fed marketplace. For example, they may have more than one group of animals in grazing rotations. The first group, either for milk or meat gain, eats only the choicest fresh grass daily and then is moved to another paddock. A second group—say of dry ewes or yearling heifers, which do not need the choicest grass (and may in fact put on undesired weight by eating it)—follows the meat- or milk-producing livestock. In some cases, even a third group of animals may be brought into play to clean up what grass and weeds remain, such as draft horses in seasons when they are not being worked hard. Or in some intensive grazing practices, a hundred lambs or more might be turned into a paddock of only one-fourth an acre but moved twice a day. The kind of grazing operation we are after on our place, with sheep and beef cattle, is not nearly that intense.

There might not be The Right Way in pasture farming, which is why it is so fascinating and why pasture farmers like to argue so much; but one way or another, this kind of husbandry allows a farmer of whatever size to remain flexible, ready to take on any new weather or market reality in this crazy world of constant change. For everything that nature and the politicians can throw at us, pasture farmers have an answer that will not demand a huge new investment of money. As an underdog all my life, my favorite old saying (it kept me alive in an all boys' high school) is: "He who fights and runs away, lives to fight another day." Applied to farmers, always underdogs to nature and political power, the saying should be: "He who finds another way, lives to eat another day."

I did not start out to establish a pasture farm. I simply wanted to enjoy myself while proving that a small farm could be a viable economic enterprise no matter what the Wise Men of Economics were saying. I came to pasture farming as a practicality. I wanted to continue my work and experiments even into the days of old age. (I hope to be lucky enough to drop dead, literally, as I work in the fields, to give the buzzards something to eat, not mildew away in some retirement village and then have my dead body paraded around funeral homes and cemeteries.) Pasture farming, I realized, was a way to produce food without so much time, sweat, or hemorrhoid attacks from riding tractors all day long. Reading Masanobu Fukuoka's classic *One-Straw Revolution*, I realized I was not alone. Wrote Fukuoka: "I was heading . . . toward a do-nothing method. The usual way to go about developing a method . . . results in making a farmer busier. My way was opposite. I was aiming at a pleasant, natural way of farming which results in making work easier instead of harder. . . . I ultimately reached the conclusion that there was no need to plow, no need to apply fertilizer, no need to make compost, no need to use insecticide. When you get right down to it, there are few agriculture practices that are really necessary."

As a young man milking a hundred cows with my father and brother, I often considered the irony of our situation. While we worked our butts off, the cows loafed in the shade. I imagined them remarking to each other about what lunatics humans must be. We were making thousands upon thousands of bales of hay and straw in summer heat and then fighting our way through adverse weather and machinery repairs to plant and harvest hundreds of acres of grain. The cows, like queens, merely waited in comfort for someone to bring food to them daily. We wouldn't even trust them to chew their food properly but milled it into a powder so that it would slide through their guts faster. Gigantic mountains of manure piled up, which we, poor stupid humans, had to haul out to the fields. It was as if cows did not have legs or teeth but were merely blobs of flesh into which we shoveled feed and from which we extracted milk. To complete the lunacy, the market price of milk was below the cost of production if we paid ourselves living wages, so we were slowly going broke.

I dreamed of moving our northern Ohio operation to Kentucky, or at least southern Ohio, where surely the cows could feed *themselves* for a longer period of time in summer and where the land was quite a bit cheaper. The time was the 1950s, and I was much influenced by the stream of articles and books on grassland farming coming out then. But such a move was out of the question for many reasons. We were much too locked into our community and into so-called progressive tractor farming with borrowed money to make a change. The notion that even in northern Ohio we might be able to get nine months or more of grazing never occurred to me. I was a son of the Corn Belt. I worshipped corn and giant hay barns. Eventually, our dairy declined even though it was big-time farming for those days and was modeled on the best knowledge the universities had. The whole project was very sad, dashing my father's dreams of having a family farm big enough to support his family and his sons and their families.

For awhile, as a lost soul, I lived in Minnesota, where I became a hired man in an even worse climate than Ohio's for my dream of pasture farming. But I learned, much to my surprise, that in the North nature packs into a shorter growing season a bigger wallop than it does in the South. A farmer from Canada told me that because of the significant variation in day length between winter and summer, he could produce more forage in one hundred days than a farmer in the southern United States could produce in three hundred. I also learned from my employer some tricky ways to substitute pasturing for grain growing and haymaking. Even in this cold climate, he grazed his dairy cows on alfalfa and corn stalks until the snows got too deep. He utilized very long stretches of single-strand, cheap electric fence to keep the cattle where he wanted them, a practice that would become critically important to dairy graziers years later. And he pastured pigs on alfalfa, bringing them to market weight on half the grain normally used. Again, I filed away these observations for later use. Someday. Someday.

I finally did get to Kentucky, still hopelessly without money enough to start any kind of farming on my own. I spent my time armchair farming while I pretended to be going to college. As Yogi Berra is supposed to have said, "you can observe an awful lot just by watching," and that's what I did. There were cattle and sheep farms all over Kentucky where the animals grazed almost year-round and were fed only a little grain, if any, before going to market, often to Europe, where consumers appreciated grass-fed beef. Grass farming was a way of life in the midsouth long before the term *grass farming* came into usage. Some of the farms were beautiful beyond any telling, with lush rolling fields of grass and clover, some of them in fescue that could be grazed all winter. The first few times I saw cattle grazing in January, I would stop my car along the road and just stare. I felt as if I were looking at something full of implications for agriculture that were not yet fully understood. But I was so beaten down with failure and lack of money that I did not trust my judgment and had no esteem for my ideas. Once more I just filed away the scenes of those lovely winter fields of grazing animals. Someday. Someday.

It took me twenty years to claw my way to the ownership of thirty-two acres in home country. It took that long because I had developed a horror of borrowing money. I had seen what borrowed money did to farmers like my father. Shakespeare gave more practical advice for farmers in two lines of poetry than a whole hay wagonload of books on modern economics: "Neither a borrower nor lender be . . . and borrowing dulls the edge of husbandry." I wanted land that was paid for. And I was sure that if borrowed money hung over me, it would distract me from my work as an independent writer, a career even more precarious than farming. Borrowing dulls the edge of literary pursuits, too. I had to submit myself to wage slavocracy, as Scott Nearing called it, until I had saved enough, even if for only such a small farm. Better thirty acres paid for than three hundred that the bank owned.

Established on my little farm, I quite by accident was gripped with the vision of pasture farming again. I had become a friend of David Kline, an Amish farmer who would later start a farm magazine that would devote many pages to grass farming. He took me to visit the farm of David Miller, a purebred Shorthorn cattle breeder. From Mr. Miller's family home, high on a hill, I could look down across about three hundred acres of magnificent grassland, dotted with cattle. It was just so beautiful. But the full significance of it only hit me when Mr. Miller told me that he did not grow one stalk of corn or any other grain or soybeans, nor did his father before him. He owned no planting and harvesting equipment. He bought a little corn, he said, but didn't think that he really needed to. I was looking at a farm where there was no annual cultivation, where the fields were all permanently in sod. It was mindnumbing. And it was commercially profitable. The paradise that I thought existed only in my mind actually existed in reality.

Shortly after that, I met Bob Evans, a lifelong farmer who also happened to turn Bob Evans Farms into a restaurant chain that made him a fortune. I should rather say that he met me. He called to say that *The Contrary Farmer* was the only book he had ever read all the way through. I was flattered and very surprised. The few people who had complimented me before were mostly philosophical types no more successful in the hard world of finance than I was. But here was a multimillionaire who had known the derision of the business world when he started selling sausage store to store telling me that what I was saying needed to be read by young commercial farmers everywhere. And, he growled, why the hell didn't I criticize Ohio State scientists even more than I did for dragging their feet on grass farming. He told me to come to see him. (Bob does not ask; he tells.) So I went to southern Ohio and met this most delightful man, shrewd, practical, gruff, impatient with stupidity, a keen mind that could spot a faker three libraries away. Not everyone got along well with him, I learned, but I loved the guy. He was demonstrating on his farm that year-round grazing, as he called it, was the way of the farming future. He grazed two thousand acres of cheap Appalachian hillsides that he had started to buy as a young man just out of the army, desperately trying to fight his way into farming on his own. What he bought was mostly a wasteland of gullies, poverty grass, and multiflora rose. "They said I was crazy to do that, too, just like I was crazy to start in the sausage business," he told me. His "wasteland" is now an absolute paradise of grass, trees, and cattle. It had taken tons of lime to make such lush grass and legumes grow, and an old bulldozer, which he continued to run himself, even in his seventies, to scrape away the jungle of multiflora rose and fill the gullies. The cattle, grazing year-round, converted the grass into meat and spread their manure to keep the pastures fertile.

I came back home and started to turn our place into a little grass farm. The first priority was to divide the land into small plots so that I could move the livestock from one to another to keep them eating the most nutritious and freshest diet possible and at the same time keep the pastures in good vigor and fertility. Although that sounds easy, and is in a way, it requires years of experience on a particular farm to know what size to make the plots, or paddocks, as graziers call them, for the number of animals being grazed, and to know what grasses and clovers are best adapted to the particular soil of each paddock. Actually, I don't think anyone has, or can, come up with a formula that fits all occasions, although plenty of experts have tried. Even when a grazier thinks he has it figured out on a particular farm, weather, or some new improvement in animal breeds or in grazing plants, or a change in market demand, can produce a new set of circumstances. The first law of farming is *every year is different.* As I experimented, I thought of myself as a sort of music conductor. I directed the plants on the various plots as if they were sections of an orchestra from which I wanted a particular sound at a particular time while I moved my animals from plot to plot as if they were a marching band, all to draw a gloriously harmonious song of the farm from both plants and animals.

I had fifteen acres with which to experiment. My first decision went against normal grazier practice. I contrarily did not want to use electric fencing, which most graziers prefer because it is relatively cheap and makes changing sizes of paddocks easy. Electric fencing is almost indispensable to strip-grazing, where the fence is moved frequently. But I had acquired some good used woven wire fencing and posts for free when highway workers put in a new fence along the interstate not far away, and I meant to use it.

I divided the fifteen acres only gradually. I wanted to be sure my woven wire permanent fences were in the right places, because changing them would have been far too much work. I already had a permanent perimeter fence of woven wire around the whole property, which I think is a good idea even when electric fencing is used for cross-fencing. My original plan called for fourteen acres in pasture and one acre of corn, the corn acre rotated from place to place on the level paddocks, never on the hilly ones.

I started erecting the cross-fencing only after considerable observation. Fences influence where livestock make their customary paths across the fields, and a fence up and down hills will usually mean cattle and sheep wearing paths up and down the hills too. These paths can turn into gullies in time. Wherever possible, I positioned the cross-fences at a right angle to the slope of the hills. The fence line itself then slowed water flowing downhill, and the livestock, sidling along the fence line, made paths that were at right angles to the flow of runoff water.

The first cross-fence divided the whole fifteen acres into two fields, one of about eight acres and the other of about seven. I didn't divide the whole into two equal parts because I wanted the fence to follow the rim of the hillside that runs across the fields, and that rim wasn't exactly in the center of the property. The result was containment of the hilliest part of the pasture into just two paddocks, with the other paddocks enclosing more or less level land. (Best to refer to the diagram as you read this.) Then I fenced off the lower flat acreage from the hillside and finally another fence line separated the upper flat acreage from the hills. At this point, I had in effect made one field enclosing the creek and brushy pasture, one level field of bottom land, two fields that were mostly hillside, and two more flat fields above the hill slope. A little later I further divided the bottom land into two paddocks.

I left the fencing at that stage for a couple of years to see how it would work. I immediately enjoyed unexpected benefits. I had previously been cultivating the flat bottom land next to the creek. That acreage flooded regularly, but, true son of the Corn Belt, I stubbornly had gone on trying to grow annual crops on it because it was such good soil. The decision to turn this land into permanent pasture relieved me of flood worries. Also, being the richest soil on the property, it proved to be the most productive pasture, even when it did flood temporarily, and more profitable than grain. Finally, turning it

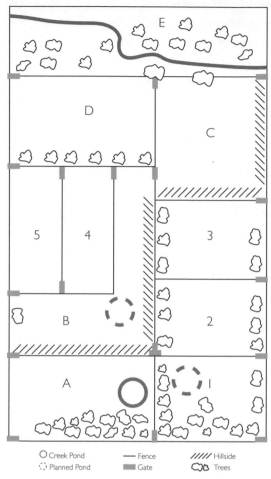

O Creek Pond   — Fence   ///// Hillside
⊙ Planned Pond   ▪ Gate   ⊖ Trees

Lettered paddocks are permanent pasture
Numbered paddocks are cultivated in rotation once every five years

to pasture allowed a wild bluegrass that grew along the creek to move back across the field. This grass is a minor wonder. It grows even in November, and at the slightest warm spell in March, it starts sending out a few green blades. On December 22, 2002, as I was writing this, the sheep were still grazing that grass, which came on abundantly after the rains of late fall. Bob Evans thinks bluegrass is a weed that graziers should get rid of. We argue about that. He wins the argument, but I keep my bluegrass.

Using agronomic averages for this area, an acre here should carry five sheep or a cow and calf. So I figured my fifteen acres should support at least twenty ewes and forty lambs, the lambs coming in April and sold in December, plus a couple of cows and their calves, the calves, too, going to market in December. Plus a small flock of chickens running loose on the plot closest to the coop. I believe that with the increase in fertility and therefore growth of pasture grass that comes with continual grass farming, our pastures will eventually surpass the carrying capacity our land is rated for, and the increasing verdancy of grass and clover are proving that assumption. No one knows what awesome production might occur in this rich midwestern soil if it is committed to grass farming for a century or two. I always have to laugh at the Morrill plots at the University of Illinois, which were planted to corn every year for over a century and declined in yields only very slowly even without fertilizers. The agronomists thought that was remarkable. If that experiment had been conducted with grass and clover in a managed grazing system for over a century, fertility would have increased to where, as in the prairie days of Illinois, one could hardly ride a horse through it and would need an Abrams tank equipped with a scrap metal grinder to mow it. You can use your own calculations and situation to see how my numbers pencil out for you. It sure beats corn and soybeans any way you figure.

I grazed with these seven paddocks of a little over two acres each for a couple of years. I was reading everything I could find on rotational grazing and was visiting grass farms when I had time. Most of the instruction I read and heard focused on the number of animals per paddock and the timing of their moves from one paddock to another. Attention seemed to be fixed on these two variables and not so much on the different kinds of pasture plants for different kinds of situations. After three dry years followed by our worst drought in history, I decided that information about different forages was more vital to me than time of movement or density of grazing stock.

I was partly wrong. I needed more paddocks, all right, or fewer animals, to allow pasture plants to regrow properly before pasturing them again. But I was correct in that what I needed first of all was more paddocks to grow different forages that withstood drought and winter snow. So I divided the field on the west half of the pasture into three paddocks and the bottom field not prone to flooding into two. That gave me ten paddocks altogether, five of about two acres, each permanently in grass and legumes, and five of about one acre, upon which I could experiment with rotating corn and other grains with clovers. Each of these one-acre plots would be in corn once every five years, and for the other four years they would be used for emergency pasture and for hay.

With the acreage divided into enough plots to grow all the different forages I thought I needed for year-round grazing and drought emergencies, I went through a trial-and-error period in experimenting with various grasses and legumes. That trial and error continues, but so far here's what I rely on. On the permanent grass pastures, bluegrass (sorry, Bob), ryegrass, white clover, and tall fescue carry the load, so to speak. On the other five paddocks I keep trying different combinations of forages: corn, oats, red clover, ladino clover, alfalfa, timothy, oats, and wheat. Right now, in 2003, the ten paddocks contain the plants listed below plus assorted weeds. Letters represent the permanent pastures and numbers represent the temporary ones (see diagram).

A. Bluegrass, white clover, ryegrass, and fescue.

B. Mostly fescue, some bluegrass and white clover.

C. Mostly bluegrass and white clover, a little fescue.

D. White clover, ladino, ryegrass, and an especially early kind of volunteer bluegrass. This is my richest field, and I expect to experiment here with more interseedings of red and improved white clovers. Not so incidentally, the sheep began grazing this paddock on March 21 this year, ten days before our pasture season usually begins.

E. The creek pasture is all wild plants and grasses, along with brush that the animals like to browse. I have a theory that there are herbal benefits to these plants in terms of mineral supplements (especially selenium) and wormers. But I have no proof yet except common sense, which tells me that animals will stay healthier if they have a very large variety of plants to pick and choose from, at least part of the time.

The temporary paddocks in 2002, heading into 2003:

1. Red clover. Will still be red clover next year, but with timothy added.

2. Wheat in declining red clover. Will be corn in 2003, fall-seeded to oats for winter pasture.

3. Alfalfa and timothy. The alfalfa is in its fourth year and declining, so I winter-seeded red clover and timothy into it in the winter. The clover and grass did not show up until late summer of 2003, but the stand was good then, convincing me that I can go from one legume hay crop to another without any cultivation whatsoever.

4. Corn interseeded with pumpkins in June. In spring of 2003 I seeded it into oats and ladino clover.

5. Red clover in 2002. Will be red clover next year, with timothy and ryegrass added.

You can see how the rotation goes. If I start with corn: (1) corn; (2) oats/red clover, ladino or alfalfa; (3) red clover, ladino or alfalfa, possibly interseeded to timothy; (4) red clover, ladino or alfalfa with timothy; (5) wheat or oats in declining legumes. If I wanted to, I could skip the wheat or oats and sow new red clover into any of the other legumes. New alfalfa seeding in old alfalfa won't do well. Both ladino and red clovers will reseed themselves to some extent.

And then back to corn. I hedge a little when I say "red clover, ladino clover *or* alfalfa." That's because if the paddock being planted is in the lower bottom land, I prefer red clover or ladino since they grow better on the moister, heavier black clay. In the paddocks on higher ground, the soil is suitable for the longer lasting alfalfa.

The only part of this rotation that is somewhat new is interplanting oats and red clover in the corn in August, and trying out sweet corn as a grazing forage. In later chapters I explain the advantages of these practices as well as details of how I try to marshal all the grass and legumes in my grazing regime. Right now, I want to march through the year, 2003 to be exact, to show how the rotations work out in practice.

April 1 is the true New Year's Day on our farm, because that is generally the beginning of the new grass. Sometimes there is some new pasture in March, too. In 2003 we started grazing on March 21. At any rate, the animals have the run of all the paddocks from winter into spring, eating stockpiled grass in winter and nibbling wherever they can find a new blade of grass or leaf of legume in spring. Graziers on larger operations often strip-graze their stockpiled winter pastures, but we haven't found that necessary yet.

We try to schedule breeding times so that the lambs and calves are coming in April. Pasture lambing has some drawbacks when a ewe and lamb need personal attention, but in general it works better for us even though we may pen ewes and lambs in the barn for a day or two after birth.

Turning the livestock out to graze this early violates an old adage of farming that I now believe is a myth. The adage claims that livestock will ruin a pasture if allowed to graze it before it establishes a good stand. In the traditional way of pasturing, where the pasture is grazed continuously, that is true because the forages never have a chance to recover. In modern grass farming, the livestock will be moved off of early pasture soon enough so that the grass and clover will revive very well. In fact, I think that early grazing is good for the pasture because it precedes the time of the most vigorous seasonal growth. Slowing down that vigorous growth keeps the grass from maturing beyond the best nutritional stage before the animals can consume it, or before hay can be made from it, or before I have time to clip it. I think early grazing is especially important to slow the growth of bluegrass and ryegrass and especially the tall fescue to keep them from shading out the white clover, which I consider my most important grazing plant.

As soon as the sod dries out enough after late winter thaw so the hooves of the animals don't sink in too much (called pugging or puddling), the livestock have the run of all our paddocks. I have learned that even legumes being reserved for a first cutting of hay can be pastured at this time in a pinch (if I am running out of hay), but only a little. The delay means haymaking starts later, but that can be a good thing, since rainy weather almost always plagues the earliest hay crop. The animals mostly nibble on ryegrass and bluegrass in the permanent pastures. Around April 1 the animals can find grazing enough so that they show little desire for supplemental hay.

On Paddock 4, I planted oats and ladino clover in late April.

About May 1, with the grass growing strongly, I penned the livestock in Paddock A with access during the day to the wheat in Paddock 2. There's a pond in Paddock A for the animals to drink from, and it is close to the barn. Being close to the barn and house and convenient to our frequent comings and goings, the lambs are a little more protected from predators.

I had planned to plant corn in Paddock 2 in May, but rainy weather made that impossible in 2003. There was an advantage in that the livestock could keep the wheat down through all of May—extra pasture I had not planned on. I did not get the corn planted until June, but disking was all the soil preparation needed. I grouched and grumbled about the late corn, which made only a spotty stand, but turned the problem into an advantage. I had always wanted to try broadcast oats and clover in corn in August and now I figured the thin stand of corn would allow enough sunlight through to make that work. Did it ever. The oats and clover formed a jungle of winter grazing along with the cornstalks and corn grain I do not harvest for the pigs and chickens. I told visitors I planted the corn thin on purpose.

About May 20, I turned the livestock into Paddock B and mowed the fescue threatening to shade out the white clover in Paddock A.

As weather allowed in June, I made hay in Paddocks 1, 3, and 5. On the permanent pasture paddocks, I clipped the grass after the animals had grazed it down suitably to control weeds and keep the grass in a vegetative, nutritious condition.

I switched the livestock to Paddocks C and E about June 10. About June 20, they went on to Paddock D, even though there was plenty of grass left in C because of abundant rains. I clipped C as I had A and B after moving the animals off. In Paddock D, as in C, the livestock also have access to E, where the creek runs, for water and their "medicinal" plants. Then in July the animals went back to Paddock A and subsequently B, then C, and by late August into D. And then I repeated the rotation until December, when the animals have access to all the permanent pasture paddocks. Because of abundant rains, I had no need to turn them into any hay paddocks in August, as I have had to do in previous dry years, so I am going into winter with lots of stockpiled red clover and alfalfa for December, January, and, I hope, February.

The precise timing of these rotations depends entirely on the weather. In the blessed summer of 2003, rains were so abundant that the cool season grasses kept on growing through the so-called late summer slump season. I learned a valuable lesson. With ample rain, there is no summer slump. The rains made a joke out of all those grazing graphs in instructional manuals that show a big upward curve of plentiful grass in early summer and a big downward curve in late summer.

I made a second cutting of hay from Paddocks 1, 2, 3, and 5 between the rains of July and early August. I took a first cutting off the new seedling ladino in Paddock 4.

In August, I broadcast the oats and red clover in the corn, as mentioned above.

By early November I got the corn in Paddock 2 harvested and cribbed. (Normally this is an October job, but wet weather had held up planting until June.) During the winter the corn fodder, plus any stray ears that I missed or left deliberately for the livestock, make good roughage with the interplanted oats.

In December and January, the animals move onto the stockpiled alfalfa and clover in Paddocks 1, 3, 4, and 5 as needed. Over the course of the growing season, I made little stacks of hay for them to self-feed from, too. I don't put much hay in the barn anymore, just enough for feeding during thaws in late winter when the animals would pug the soil too much if allowed to walk on it or after shearing, when for about ten days the sheep don't have much protection from the snow and rain.

During September, legumes need to have a chance to store fertility in the roots for next year's growth, so the paddock grazed at that time should be the one that is in decline and scheduled for reseeding the next year. Once mid-October comes, legumes can be grazed to the ground without hurting them. More on that in the chapter on legumes.

In late October we turn the ram in with the ewes. In November and December we sell lambs and calves.

In late winter, after the stockpiled oats and clovers run out, we have to rely on hay. Judging by the experience of others who stockpiled oats and were able to graze them through February and even early March, I hope we will come very close to our goal: year-round grazing.

In early spring we will replace or repair fences where necessary. The soil is moist and soft at this time, making it easier to dig post holes or drive posts. At this time I will also spread fertilizer on whichever paddock is in its second year of being hayed. Eventually, I think I will be able to stop this practice. Soil testing will tell. Hay crops take lots of potassium out of the soil, and I am not sure yet that my soil has enough fertility to make up for that loss. I replace it with muriate of potash. I would prefer a more organic potash but none is conveniently or economically available here. I use leaf compost from our village whenever I can get enough. When I have spread it a couple of inches thick over a paddock, the response has been tremendous. If I could afford to put a half foot of compost on the whole fifteen acres every year, I quail at what the result might be. A century of that might produce night crawlers the size of crocodiles. Sheep would grow as big as horses, and horses big as elephants. Cows that could not learn to run fast would be overwhelmed by the lightning-swift regrowth of ladino clover after rain and be suffocated in their tracks.

Even with only occasional applications of compost, plus the yearly clippings of surplus pasturage rotting into the soil, plus the manure of grazing animals, that would surely mean . . . well, let's be serious now. Farmers would for the first time in history make a little real money. Enough to keep on farming, anyway.

3

Some Commercial Grass Farms

Commercial grass farms can be of almost any size. If the grazier were raising high-priced animals like some of the new miniature cattle breeds or, perhaps in the future, milking cows that carry inserted genes to make the milk especially beneficial for diabetics (which is being done, by the way), the operation could be as small as a few acres and make good money. Or a pasture farm could be thousands of acres, like a western ranch. But the kind of grazing I am mainly describing here is moderate in size, in the range of fifty to five hundred acres.

There is also more than one kind of commercial farmer within this class. There are farmers who consider themselves pure graziers and, especially in dairying, those who call their farming "grass-based." A "pure" grazier might grow no grain or hay, relying almost entirely on pasture and buying small amounts of supplemental feed if necessary. Grass-based dairymen are more like traditional farmers. They have cut the amount of grain they feed drastically in favor of rotational grazing. They have also reduced the amount of hay somewhat, but they still grow some corn and hay. There is much debate between the two, which is why I keep warning readers not to expect too much agreement between grass farmers. We are a farming society in transition.

I need to point out that pasture farmers should be further divided between veterans and more or less beginners.

That's another reason why we like to argue. A farm that has been in permanent pastures for twenty years is a whole different animal than a farm coming out of decades of commercial corn and freshly seeded to grass. If only there were true pasture farms that have been in operation for one hundred years like there are grain farms, then we could make some truly informed comparisons. Here are some examples to show the wide diversity.

Dairy Products with No Grain

At the Ohio Environmental Food and Fiber Association conference last spring, I stopped by a booth offering samples of cheese (I hardly ever pass up free food) and was struck by how much I liked it. I am not particularly a cheese eater. The cheese was Meadow Maid from Ohio Farm Direct, a new business operated by Doug and Pam Daniels, who were handing out the samples. Their dairy farm has been in the family for seven generations. When Doug announced that he was going to quit feeding grain, he says his father wasn't exactly thrilled. "Yes, we lost production, maybe 20 percent," he says, "but the nutritional value of the cheese over cheeses from grain-fed milk shows no loss but gain." He sells his cheese on the basis of that increase in nutritional value and will bend your ear on that subject for as long as you care to listen. He currently has enough demand for his cheese that he is offering five dairy farmers a premium *higher than the premium producers are getting for organic milk in Ohio.* He is looking for more producers, but it is difficult to find dairy farmers who believe they can produce milk profitably without any grain.

When I reiterated later on the phone how much I liked his cheese, he responded: "The answer might be that because our cheese comes from grass-fed milk without grain feeding, it is exceptionally rich in omega-3 fatty acids, as is meat from totally grass-fed beef. Your body might be hungering for more omega-3." (Much nutritional evidence suggests that omega-3 fatty acids can help lower the risk of cardiovascular disease, cancer, depression, allergies, autoimmune disorders, obesity, and diabetes.)

What Mark and Debbie Apple told me about their grass farm in Indiana would raise the eyebrows of a mummy. They violated the three main articles of dairying faith. They fed their cows *no grain.* They milked only *once a day.* Their cows were Dutch Belted, not Holsteins.

When I was milking a hundred Holsteins forty years ago, that kind of operation would have been used as an example of how to go broke in a hurry. I asked Debbie how much their production had gone down on once-a-day milking and no grain. Fifteen percent, she answered readily. But because they were selling milk retail to their own customers at their own price, she explained, they actually were making more net profit than when they fed corn and milked twice a day. And of course they were working less. Interestingly, she said that though the volume of milk declined by 15 percent, the amount of cream in it did not. Evidently, what they were mostly losing was water.

Do the cows dry up early when milked only once a day? (Some confinement dairies milk three times a day.) No, she said. Another article of dairying faith down the drain.

In fact, says Debbie, though it takes a year or so to condition the cows to an all-grass diet, once that is accomplished, the cows take on a sheen and vitality they did not exhibit on grain. "The black part of their hair just glows," she says. They are milking only eight cows during their experimental phase but plan to add fifteen soon.

One of the keys to their success, they explained, was their ability to sell their milk raw to the public, a practice barred in many states. The law usually allows farm families to drink their own raw milk (as if it could stop them), but not to sell it to other families. To me, that's an indication that the giant dairy cooperatives are using pasteurization as an excuse to monopolize the milk market. (There is some startling evidence that raw milk has more nutritional value than pasteurized, homogenized milk.) Anyway, following a gambit that Wisconsin farmers have been using, Mark and Debbie sold shares in their cows to customers. The customers thereby became part owners of the cows and so could legally drink the raw milk from them. "Share milking," as it is called, was challenged in Indiana, and it took a lengthy battle with the bureaucracy and finally the help of the governor personally before Mark and Debbie were allowed to proceed. Because customers said their milk tasted so good, and because knowledgeable people are looking for "grass-fed" milk, the Apples had no trouble selling their product. In fact, the local chapter of the Slow Food Association, which encourages quality, healthful food, has been so enthusiastic over the Apples' milk that it has all but adopted the little dairy as a prime example of where food production could go in the future.

Milk with Reduced Grain but No Corn Silage

I have not had a chance to visit the no-grain dairy farms mentioned above. But I have visited several times the farms of Nathan and Kristine Weaver and of Tim and Katie Kline, neighbors in east central Ohio. They offer an absorbing example of two somewhat different approaches to grass farming.

Nathan milks about forty cows. He grows no corn or corn silage, but he buys a little grain and feeds hay along with grazing. He milks seasonally—that is, for ten months, with about two months off in winter. He feeds a daily ration of purchased grain of about ten pounds per cow. Conventional farms feed twice that amount.

Nathan grazes his forty cows and about half that many young stock on only sixty-five acres of permanent pasture, along with some annual pasture and hay. His pasture crops are mainly improved varieties of ryegrass, white clover, and fescue, with some bluegrass. He is experimenting with a paddock of mixed turnips and sorghum-sudan grass to stretch the grazing season into winter. He grows alfalfa for hay. He gets his cows bred all within the same month, or nearly so, using artificial insemination, so that the cows can all be dried up at about Christmastime. The family then does not have to milk again until the cows freshen in March. He does this not with synchronizing drugs that bring the cows into heat on demand, but by taking advantage of the cows' natural heat cycles. Not only does seasonal dairying mean that the family has a two-month vacation from milking during the coldest part of the year, but the cows are dry and so require less high-energy and protein feeds during the season when there is little or no pasture to provide it.

The pasturage is divided into paddocks with single-strand electric fencing. Nathan's father, who is about my age and who helps out with the milking, supplied me with a forked stick when I walked the pastures, making it easy for me to push down the fence and step over it without the risk of getting shocked from having to high-step over it. I highly recommend forked sticks for pasture walks. The paddocks number fifteen but are subdivided, with water tanks strategically placed where they can serve more than one paddock whenever possible. Water is delivered to the tanks by plastic pipe from a well near the high point of the farm. The water is pumped by a windmill or, when necessary, a two-horsepower engine into a cistern, from which it flows by gravity to the stock tanks in the fields. "We don't use the creek. The cows pre-fer the well water. We worry about possible pollution from upstream."

As Nathan says, he is never quite sure when he might need to re-arrange the size and shape of paddocks. Different circumstances, changing weather, and varying numbers of cows and calves constantly demand different arrangements, easy to do with single-strand electric. He moves the cows to fresh pasture every twelve hours. Young stock and calves move onto pasture vacated by the milk cows. Workhorses follow the calves to clean up the paddock before it is allowed to regrow for the milk cows.

When I was there in early August, usually the low point in sum-mer pasture growth, the ryegrass, bluegrass, and white clover were magnificent. This was due in part to abundant rains that year, but also to the buildup of nutrients in the soil that followed years of managed grazing on permanent pastures. The pastures were supporting the equivalent of more than one cow unit per acre, counting the calves and young stock plus the horses. There was no shortage of pasture but actually a surplus—in August!

I asked Nathan if he thought he'd ever grow corn again. He wasn't sure, but if he did, it would be for somewhat the same reason that he had quit. The amount of corn he feeds is so small that he figured it was more economical to buy it than to spend the time and money to raise it. But by the same token, if he did decide to grow corn again, he would need so little acreage that it would not interfere with his pasture operation even on his small farm. His reasoning bears repeating. "First of all, we have proven to our own satisfaction that the protein and en-ergy from good pasture exceeds that of the typical confinement dairy ration which consists mostly of corn and alfalfa. With pasture, a ton of corn per cow per year is the most I need to feed. So, if I were to grow that myself, my land, especially after it has been building nutrients as pasture for several years, is capable of producing corn at 200 bushels per acre. That's enough corn in my operation to feed six cows. So I would need only seven acres to produce all the corn I need."

Nathan does not feed corn silage, which is an almost unthink-able departure from traditional dairying but which amused me be-cause I agree with him. "Remove the grain from ensilage and you have an extremely low-quality feed," he once wrote to me. "Corn silage has come into favor because it is easy to produce although expensive compared to grazing. On the other hand, it takes skill to produce a nearly balanced dairy ration on grazing alone. We graziers get paid for being smart."

Other gems from Nathan:

> The advantages of grass farming over confinement dairying go beyond herd health benefits, important as those benefits are. The cost of production is much lower. It is not because of lower tillage needs or even less expense for seed and fertilizer. The main thing is that grass farming can mean the end of harvesting and storing and processing and feeding and hauling the manure, the things that take the most money and labor.
>
> What we really need now is the right kind of cow for grazing. That's the missing link to progress in grass dairying. Cows have been bred for over a century to produce lots of milk on a high grain ration. We need to start breeding for thriftiness and high milk production on pasture. Grazing cow genetics in the U.S. is a wide open field. Sometimes I think we're still lost in the woods somewhere.

He has switched from a purebred Holstein herd to cows of several breeds and crossbreeds. He uses mainly New Zealand bulls and a few milking Shorthorn bulls. "I don't think we have the proper grazing genetics in the U.S. cow herd to achieve our goal of cows that perform well on a high forage diet," he explains. He then saves calves from the cows who produce the best on grass. "I don't care what color they are."

Then he adds: "However, I do believe in order to stabilize our herd so that the genetic outcome is more predictable, we graziers will need to develop a purebred cow adapted to what we're trying to accomplish."

Milk with Reduced Grain but Plenty of Balage

Overall, Tim Kline's farm is similar to Nathan's, except that he has started raising meat goats, too, as a possible second source of income. The goats in the meantime play a key role in weed control. "They love multiflora rose and blackberry bushes," he says with a grin. "They even eat ironweeds."

He has his farm divided into an array of grazing paddocks surrounded by single-strand electric. Other fields are in hay and corn. His grazing paddocks are about two to three acres in size across which he advances the polycord to give the cows what fresh grazing they require each day. Unlike many graziers, he does not use a back wire behind the cows as they strip-graze. "Allowing the cows to walk back over what they have just grazed a day or two earlier doesn't seem to matter. And saves me time plus I don't have to pipe water everywhere. I only have water at the gate to the paddock."

He relies on Alice clover, an improved white clover, with ryegrass and tall fescue as the mainstay of his grazing. "Alice is a wonderful grazing legume, in my opinion," he says. "It endures, spreads by both root and seed, is very good at covering over bare spots in the field, produces more forage than regular white clover, yet it is much less likely to cause bloat than red clover, ladino clover or alfalfa. I keep a daily diary of milk production and which paddock the cows are on," he says. "With a good stand of Alice with the grasses, I can gross as much as $2,500 of milk per acre annually. On paddocks with lesser stands of common white clover, the gross is more like $1,900 an acre."

He frost-seeds the clover into existing sods. Once established, the Alice endures without reseeding. "In fact it seems to compete almost too well with the grasses," he says. He prefers ryegrass and tall fescue over bluegrass because the latter produces much less forage in his experience. He admits, however, that he has not had experience growing bluegrass and that new varieties are coming along that might make it compete better. "Fescue must be clipped frequently, in my opinion," he says. "Many graziers do not believe in clipping, but I do. If fescue is not clipped, the cows won't eat it as well, and it will overwhelm the clover."

Unlike Nathan, Tim believes in growing his own corn. "We have just not had much luck with purchased grain rations. The kernels of commercial corn just seem to be too hard." He smiles. "I envisage those kernels rattling around inside the cow, bouncing off the top of the rumen, and shooting out the rear looking about the same as they did when the cow ate them. We've had acidosis problems with commercial rations. I am going back to the soft corn like my father grows. You won't see cracked kernels in the manure when the cows are eating *that* corn."

Tim thinks that corn silage is a good source of energy. "And it is as close to a guaranteed crop as you can get in this part of the country where rain often makes haymaking risky." Instead of only traditional hay, he is now relying on what is called "balage." Balage is hay not yet completely dry, baled up into big round bales wrapped in plastic. "I like to make it when the forage is almost dry enough to store as hay," he says. "Then it keeps well inside the plastic. If it is wetter, it will smell sour and is not as palatable. The cows love good balage. The other advantage is that we can make hay sooner if rain is threatening and we can make a lot of it in a day with big round bales."

He prefers to feed his cows in the barn in winter weather and sometimes in hot weather. "You can do things with sheep and cattle that I would be afraid to do with milk cows," he says. "When the cows start bunching under a shade tree, I will let them go into the freestall barn and put balage out for them, especially since we have lots of balage this year. As for winter, I would never be able to sleep on a windy winter night if the cows were out in the weather like they are on some grass farms. Even on milder winter days, mud can cake on the cows' teats and cause problems. And if you feed a big bale out in the pasture, the cows will churn up mud and ruin the sod around it. Also the manure will be concentrated in that area and the cows won't eat the grass that comes up there the next spring. That's partly because this farm was abused for years before I bought it. I haven't had time yet to build up a good balance of biotic life in the soil to decompose manure quickly."

Tim also points out that there is an advantage to having some manure in the barn. "I can haul it to the spots where it is most needed rather than having it pile up under the big old pin oak tree where the cows bunch for shade. Also I think cow comfort is our God-given responsibility."

He doesn't practice seasonal dairying. "I just don't think I can afford it," he says. "I have a long way to go before I get this farm paid for. I need to milk full time. That's one of the reasons I want to grow my own corn too. I still think I can raise it cheaper than I can buy it. And I can control quality better that way."

What about just not feeding grain at all? Tim shakes his head. "I know some are trying it. Some of them aren't in business anymore, either. That may be for other reasons than not feeding corn. But I just couldn't afford the risk of losing production. And I don't want to sell milk retail. Not my style. I wouldn't have the patience to put up with a guy from the city coming in here to buy my milk and wanting to stand around for two hours talking to me because he thinks he's doing me a favor. I guess what I'm saying is that there are different ways of grass farming, and I'm trying to look at it in ways that fit the majority of dairy farmers."

A Sheep Farm with No Annual Cultivation

In the fall after grain harvest, the industrial grain land across the Corn Belt is one vast, rolling landscape of brown soil open to erosion— "brown fields," literally. The land has been plowed, chisel-plowed, or disked, then treated with herbicides and fertilizers, all in an effort to get the soil partially prepared for planting fast the following spring. But as I approached Bruce and Lisa Rickard's farm in central Ohio, I saw green. All their fields were permanently in pasture and remained remarkably summerlike with erosion-proof grasses and clovers even as winter approached.

"That's right," says Bruce. "We don't raise any annual cultivated crops. No corn. No soybeans. No plows, no chisels, no big tractors. No erosion. No herbicides, no purchased fertilizer, no synthetic hormones. We maintain continuous meadows. The sheep and cattle do most of the work of harvesting and weed control by grazing and spreading their manure for fertilizer. Planting is mostly a matter of adding forages to existing pastures by broadcasting on top of the ground. And the only time we have had to do that was when shifting a pasture from cool season alfalfa/orchard grass to warm season lespedeza/switchgrass."

The Rickards cultivate as little as possible, Bruce notes. "The only cultivation we've used is for small experimental plots of annuals like grazing maize or kale and turnips and even that involved only a light rototilling followed by cultipacking for broadcast seeding. Our rule is that if we can't hand-seed, we don't do it."

Their livestock stay on pasture year-round. The only barn is a shearing shed. In winter, the sheep eat grass and clover from fields not grazed in late summer, along with hay made from surplus pasture in early summer. "We really wouldn't make any hay except that doing so in the early part of the summer keeps the pastures lush and actually increases the total amount of forage," says Bruce. "Our goal is to graze year-round and not only get rid of the soil cultivation tools, but the mower, rake, tedder and baler too." As it is, they make hay only until July. After that, forage not needed for summer and fall grazing is stockpiled for winter.

The Rickards found that sheep thrive on good pasture without grain. Bruce and Lisa, along with their teenage children, Jesse and Hannah, currently maintain a flock of 520 ewes on about 170 acres of pastures. That's a stocking rate of about three animals per acre. "The farm could carry twice that," says Lisa, "but that would increase the risks from internal parasites."

They experimented with several breeds before settling on Polypays. Polypays do well with weaning at 90 to 120 days, comparatively late by most commercial standards, but just what the Rickards wanted. By 120 days, most of the lambs are self-weaned. They can be bred at 150 days, so the lambs can be all weaned at once at around 90 to 120 days, making allowances for the slower-developing lambs, and bred so that they lamb all within about a 34-day cycle, which is what the Rickards aim for. They tried the Clun Forest breed, noted for thriving on grass alone, but the sheep were hard to keep enclosed with electric fence, and it appeared to them that because of the breed's bulging eyes, the sheep were prone to pinkeye. They also tried Rideau Arcott to increase carcass size, but the breed's long legs and nervousness made them excellent fence jumpers. Cheviots proved a good breed for their operation, but their smaller size and longer time to lambing did not fit their marketing plan.

For persistent weed problems, the Rickards use very high grazing densities, as much as five hundred sheep on one-fourth of an acre for two or three days, feeding hay at the same time as a supplement. The sheep eat and trample the weeds, and the hay fed to them provides clover and grass seeds for reseeding.

They have tried some of the more esoteric grazing plants like chicory and grazing maize. They thought the chicory would make winter pasture, but it turned out to be quite tender to frost. They intend to keep using it, however, for grazing before frost. They see possibilities with grazing maize, too, but are not too enthusiastic. "We're reluctant to till up ground that's in a good sod," explains Lisa. "There's always the risk of crop failure, the cost in time and money for seed and tractor time, and anyway, our animals seem to do very well on our mixed grass/legume pasture, even stockpiled in winter. If it ain't broke, don't fix it." (The faint cheering you might hear in the background while reading those last two sentences is me.)

As might be expected, the Rickards have to contend with coyotes and stray dogs. Around the fifty acres used for pasture lambing, they install electric net fencing—Tensionet from Premier Sheep Supplies. Then they move the fence to go around the lamb pasture after weaning. The fence has already more than paid for itself in cutting losses.

The Rickards have developed their own client subscription list for their naturally raised lamb, fresh eggs from uncaged hens, and gourmet ground beef. They also sell regularly at the local farmers' market. What they can't sell directly goes to the commercial market at Mt. Vernon. They try to sell at times when prices are traditionally higher, as around the Muslim holy days of Ramadan, when demand for lamb increases. Buyers with Islamic backgrounds sometimes arrange to come to the farm to do their own ritual slaughtering.

The Rickards put meaningful integrity into their marketing. They are not certified organic because that would mean using no wormers. "The only way to do that would be to pen the sheep on concrete feedlots, I suppose," says Lisa. "We don't think that feedlot lambs, even if they were organic, would be an improvement over our naturally raised lamb." But because of consumer demand, they sell their lamb and beef as totally pasture raised. Not one speck of grain. Last year when bad drought limited the amount of good grass they had available for fattening lambs and steers, they bought expensive alfalfa meal, not cheap grain, to make good on their promise.

The Champion of Year-Round Grazing

I know a few people who are sort of famous, a few more who ought to be famous, but only one who is famous for the wrong reason. Bob Evans is a name instantly recognized all over the eastern half of the Midwest because of the chain of restaurants he founded that bear his name. Not so many people know that he has also always been an active farmer and has remained one since retiring from the restaurant business. He is a very special, even revolutionary, kind of farmer. He has demonstrated that a profitable beef business is possible without taking the tractor out of the barn, without, in fact, necessarily owning a barn at all. "Barns can actually be unhealthy for cattle," he says with a mischievous little grin, always delighted to challenge tradition. "They can do just fine out in the snow if they have a good windbreak." He waits for me to argue. When I don't, he adds: "And you can dynamite all those dumb silos, too."

As a matter of fact, dairymen are abandoning their tall silos, their "bankruptcy tubes," as some graziers call them. But I don't think most of them will quit making corn silage any time soon, as Bob Evans knows very well, even though cows on silage generally get so loose-bowelled that they can defecate through a key hole for an hour without stopping. Dynamiting silos is just another way for Bob to express his conviction that at least as far north as southern Ohio farmers can raise cattle year-round on pasture alone. They don't need to haul all that forage into a silo and all that manure back out to the fields. I've asked him on occasion to look over what I was writing about pasture farming. His usual comment: "Pasture farming is too soft a phrase. Say 'year-round' grazing. And say it two or three times in every paragraph."

What has become known famously as grass farming was not yet famous when Bob first started thinking about it. The idea grew on him gradually out of his experiences as a cattleman. He knew that in his southern Ohio climate there were mild winters when he was grazing almost year-round already. He began asking the right questions: What if cows were developed that could utilize pasture more economically than what he calls "those big lengthy things" that are bred to get fat on corn? What if science were to spend as much time improving forage plants as it has spent improving corn? What if farmers got smart enough to combine the right animals with the right plants so that the animals did most of the farm work themselves simply by grazing? Park all those horribly expensive tractors, plows, chisels, rippers, disks, levelers, knifers, sprayers, spreaders, dirt pounders, planters, and harvesters for good.

The vision of that future took hold of him, and since about 1992, at an age when most men have been long since retired, he has relentlessly pursued it. He has watched the decline of family farms with increasing dismay and realized with his practical business mind that the perfection of low-cost, year-round grazing could be the salvation of small farmers, especially in his Appalachian region. When he visited New Zealand and saw how farmers succeeded there with total grass farming, sometimes using no more machinery than a four-wheeler to cruise around the pastures, he was sold.

He threw himself into the task of persuading American agriculture of what New Zealand and Argentina already knew. "Our farmers are confused, they are afraid of change, they hold on to the obsolete failing practices of the past," he wrote in a form letter sent to thousands. "Worst of all they have worries about losing their farms, for they are struggling against forces beyond their control."

At first he was met with disbelief and even disdain. Occasionally, he still is. "Doesn't bother me much," he says. "I'm used to that. I had the privilege of growing up poor and learning to make my way in spite of it. If I had known in 1945 what I know now about year-round grazing, I would never have gone into the sausage and restaurant business."

No one was going to tell Bob Evans where the profits existed or did not exist in farming as long as he could add and subtract. As a boy growing up in the Great Depression, he found so many ways to add to the family income that one summer he says he actually outearned his father, who was having a tough time finding work. "One of my specialties was hanging around the produce houses where chickens were being bought and sold. Inevitably some chickens would get away. We boys got paid to catch them. I caught my share."

He grew impatient with the land grant colleges' slow response to his urgings about year-round grazing. He decided to do his own research. He conferred with grazing specialists in Oklahoma, Kansas, Missouri, Kentucky, and Ohio. He encouraged breeding programs to develop cattle particularly suited for grazing, backing the programs with his own money. He scoured the world for plants that would enhance year-round grazing. The Bob and Jewell Evans Foundation funded on-farm demonstrations, cosponsored numerous educational meetings, and helped develop printed materials on year-round grazing. He hired Ed Vollborn, a retired Extension Service worker who shared his vision, to conduct grazing experiments. Ed's brother, Fred, was already working as his farm manager.

He set aside land for his first experiment and put fifty head of his Charolais on it—"not the best type of cow for year-round grazing, but that's what we had available," he says. "The best cow for grazing hasn't been invented yet." The experimental herd not only survived but thrived almost entirely on grazing. Fifty cows were fed only eleven round bales of hay as supplemental feed to the pasture *in five years.* In addition, the cows got 20 percent protein range cubes at a rate of no more than four pounds per cow per day to supplement hay when quality was poor or when grazing was limited in late winter. The cost of supplemental hay, cubes, and minerals never exceeded thirty-five dollars per cow for a winter. In addition, the same land area, approximately 140 acres, that at first supported only the fifty cows, went on to support an additional sixty yearling cattle for four months during the summer grazing peak. The pastures, needless to say, kept getting better.

Meanwhile, Evans and the Vollborns were experimenting with new forages in ways that would make year-round grazing more practical. Tall fescue remained the main grazing plant for winter but was now teamed with such unusual forages as Red River crabgrass, kura clover, switchgrass, and Aroostook rye. The latter proved to be very hardy when planted about August 1 for late-fall and early-spring grazing. They found that triticale, a cross between rye and wheat, could be grazed through winter even into late February. They sowed it in the fall in declining hay fields, first disking lightly and then seeding into the mulch with a cheap, old Van Brunt drill.

Ed Vollborn also tried the age-old practice of grazing summer-sown turnips and found it satisfactory for wintering over dry mother cows. Young stock needed a little grain or good hay to go with the turnips, he says. "One of the last research projects I did before retiring from the Extension Service was with turnips. I was able to maintain bred heifers on turnips during the November-December period for an out-of-pocket cost per cow of 19 cents, including a little stockpiled pasture. Three-fourths turnips and one-fourth meadow regrowth to be exact. Turnips are usually too watery to feed alone."

Presently, Evans's farm is trying a new commercial application of year-round grazing—that is, raising stocker cattle for other cattlemen. Tall fescue is still the main forage. After the fescue is grazed in early summer, regrowth is stockpiled for winter. August-sown rye is grazed until December after it has established itself. The triticale is winter-grazed and lasts into February. Turnips are grazed as needed but preferably during November and December. Very little hay is fed. The stockers arrive at the farm about October 1 and get five pounds of grain per day, provided by the owners of the cattle. That's about half the amount fed in a conventional program. The stockers go to the feedlot April 1. Several other custom options are being offered, such as fall or spring backgrounding programs and heifer-growing programs.

Ed Vollborn says that the best application of year-round grazing he has experienced so far was with the cow/calf herd on his own family's farm. There he follows a schedule of early calving, weaning before the weather gets bad, and sending the dry bred cows back out on winter fescue. "The cows' nutritional needs are low in early pregnancy and the stockpiled fescue is enough," he says. "Through spring and early summer on good pasture, they shape up so well that they wean 600-pound calves in December. We have had to supplement the 25-cow herd's year-round grazing with *only one big round bale in three years.*"

Grass Farming Attracts Innovative Minds

I could describe many other examples of pasture farming, but readers get the idea. I can't resist just two more choice tidbits that aren't so much about grazing itself, but how grazing, as an innovative system, attracts innovative people. At a meeting a few years ago, I met a young couple, Matt and Tracy Cunningham. Matt was one of the speakers. He recounted how he and Tracy had left secure jobs in the city to live on a neglected farm that was in their family. It was mostly scraggly hills overgrown with multiflora rose, as was most of the land around it. Matt got some sheep, eleven to be exact, and bought a tractor and rotary mower to see what he could do with the thorny devil bushes. He had not been a farmer and so looked upon the whole situation with a kind of freshness and optimism that would have escaped a veteran farmer confronted with acres of thorns. To his surprise, he found that all he had to do to reclaim the land was turn the sheep loose on it and mow the hell out of the multiflora. In no time, almost miraculously it seemed to him, the native grasses and clovers started coming back. Eventually, he had most of the farm in profitable pasture. Neighbors, mostly absentee owners, took notice as they drove up and down the roads, waiting for their land to grow in value as possible residential property. Matt, noting the seemingly limitless stretch of thorn jungle, took notice also. When an admiring neighbor complimented him on his pastures, Matt replied that he could do the same with the neighbor's land and that in payment he asked only to let his flocks graze there and keep the pasture beautiful. The neighbor thought that was an excellent idea. He wasn't getting income from the land anyway, and turning it into pasture would enhance its value, especially as prospective residential property. The deal was struck and worked out fine. Then another neighbor and yet another wanted the same contract. With free pasture, Matt and Tracy's flock increased to about 400 ewes by 2002, and, Matt says, there is still enough free land available "for our flock to grow to between 1,000 and 2,000 ewes."

Richard Gilbert, whom I will mention again in later chapters, began raising sheep in southern Ohio about five years ago. He, too, knew little about what he was getting into but decided that pasture farming just made too much sense not to be the type of agriculture he would follow. But what kind of livestock would he graze? He made a shrewd decision. He would raise hair sheep, something veteran farmers hardly knew existed and would mostly think a hairbrained idea. His reasoning was that there seemed to be many people interested in raising sheep on small farms, but shearers were becoming increasingly difficult to find. Hair sheep didn't have to be sheared. He chose Katahdins, a hair sheep acclimated to the Ohio climate.

Sure enough, he was right. Shearers continued to decrease in number and would-be sheep owners clamored for hair sheep. Richard could sell most of his Katahdins as breeding stock—that is, higher than market lamb prices. What's more, the meat was milder—less muttony—than from most breeds of wool sheep. Lamb fanciers, especially those with Middle Eastern backgrounds, preferred the meat from hair sheep. All Richard had to do was allow for time to convince the commercial buyers that Katahdins aren't goats, and that, too, is happening.

His pastures are remarkable. He keeps a breeding herd of about fifty ewes on fourteen acres! In order to do that, he has adopted some novel practices, such as keeping the ewes on hay for two months after weaning the lambs in high summer, so the lambs get all the good pasture to fatten up on. He has experimented with a variety of old, new, and improved varieties of clovers and grasses, including lespedeza, which works fine on his land. But after five years he says he has experienced a near-religious conversion. It is white clover that has responded the best to his various pasture renovations on land that was almost entirely in tall fescue when he bought it. "I see now that white clover is the basis of pasture farming in this climate," he says. He likes the improved variety Will and one from New Zealand called Kopu II, but notes that native white clover blooms and reseeds better.

Although raised "on the beach in south Florida," Richard Gilbert (who has become a close friend) is proving himself to be a genuine farmer. How do I know? He is now saying what all farmers say the world over: "All I need is a few more acres."

4

A Pasture Garden

Calling a pasture a garden is a bit of a stretch, but it is an appropriate label for a small backyard grass farm. Gardening is a relaxing and more or less recreational pastime for its devotees, while commercial horticultural pursuits can be worrisome and fraught with financial risk. In the same way, pasture gardening is a relaxing, low-cost, and low-labor avocation, while commercial pasture farming must answer to the stern dictums of the marketplace. Obviously, grass farming is easier and simpler on very small acreages with very small numbers of animals. That's why it fits well with the goals and schedules of what I like to call garden farmers, those who have another job or career and look to their home pursuits as a form of enjoyment that at the same time provide good food and meaningful work. On many days a pasture garden requires nothing more than sitting at the breakfast table or reclining in a hammock while watching the animals graze. And what the pasture gardener is learning, from the hammock, he or she may someday turn into a commercial farming venture.

A second advantage of the very small or backyard pasture system is the ease with which it can be combined with traditional gardening and traditional lawn management. Lawn and garden tools are all one needs for the small pasture. The conventional garden and even parts of the lawn can become grazing paddocks. After all, before the lawn mower became popular in the latter part of the nineteenth century, the first lawns, or parks, as they were called, were kept trimmed by sheep and cows.

We frequently visit an upscale suburban enclave where we have family ties near Cleveland, Ohio. Becoming a grass farmer, I see the world through green-tinted glasses now. It now strikes me, whenever I visit this subdivision, how much good pasture is going to waste there. My sheep would think they had died and gone to heaven if they had access to those gorgeous lawns, green *even in January*. Listening to these homeowners talk about how they manage their lawns, I realize that they know more about managed grazing than most farmers do. All they really need to do is add white clover to their bluegrass and fescue lawn mixes and mow less often and they'd have pasture any dairy farmer would envy.

Statistics claim that there are thirty million acres of lawns in the United States, not counting the miles and miles of roadsides that the highway department mows and the miles of county roadsides that rural lawn fanatics keep as trim as newly shorn ewes. By mowing lawns, says research from the University of Sweden, we burn up eight hundred *million* gallons of fuel. That's over $1.2 billion dollars at today's prices. For this our soldiers die in Iraq.

Thinking like a grass farmer, I did a little calculating. If homeowners added white clover to their lawns, which some do, I presume well-fertilized, herbicided, watered, and otherwise coddled grass would produce about eight tons of dry forage per acre per year, because that's what good farmland can do. That would mean that some 240 million tons of what could be good hay are not being produced. Multiplication yields some stunning milk or meat production figures. And remember that "hay" in the form of grass clippings is now going to landfills or composting facilities at colossal expense.

Bear with me. According to my favorite grass-farming book, *Fertility Pastures* by Newman Turner, written in the 1950s, Mr. Turner's Jersey cows maintained an annual herd average of 750 gallons of milk per cow per year when he fed them exclusively on grass silage for the four months of winter. (The other eight months they grazed.) Each cow ate about 7 tons of silage over the winter, so on a yearly basis each cow would need about 30 tons of silage. If I figure that an acre would produce 20 tons of silage (if you think that's wrong, use your own figure), that would mean that 30 million acres would produce 600 million tons of silage, right? If each cow ate about 30 tons per year, following Mr. Turner's feed rations, 600 million tons would keep about 20 million Jerseys for a year. If each cow produced 750 gallons of milk per year, that would be 15 billion gallons, right? At a retail price of two dollars a gallon, that's $30 billion dollars. Add to that the $800 million paid out for fuel, plus who knows how many billions spent on lawn mowers. This is the kind of wild thinking that overcomes a grass farmer wearing green-tinted glasses.

A little more seriously, there are many homeowners who have huge lawns they don't need and are tired of mowing. (The old book *Five Acres and Independence*, written in 1935, gives as one of the reasons for failure at farming "too large an area given to lawns.") Or they have grass and brush land because they just like the open space it provides. Or, more often, they have a few acres where they keep a riding horse or other animals. But instead of managing their acreage to produce most of the feed for their animals, they buy hay and let their land become little more than an exercise lot.

Here's an example of what can be done with five acres. The names are obviously made up, and some of the details are made up, or rather they are a composite of several real homesteads headed toward the complete "garden pasture."

The Pleasures of Pasture Gardening: The Uncut Version

George and Mabel Cottager bought five acres of land in rural central Indiana because they wanted to live, and wanted their children to live, the happy and meaningful life they had experienced growing up on their parents' farms. Because of the wolfish nature of farm economics, their homeplaces had been gobbled up by industrial grain producers, sending people like George and Mabel to Indianapolis to find jobs. Now they were coming back home—buying back some of what they considered their birthright. The five acres that they purchased were part of a 160-acre farm that had gone on the block after its farmer-owner died. As has become customary, because of the demand for rural residences with a few acres, the farm was split up into relatively small parcels instead of being sold as one unit. Marvin Grabacre, who had planned to buy the entire 160 acres to add to the 3,000 he already owned, was furious. He knew that most likely the total price of the parcels sold to various home buyers would be higher than he could afford to pay for the whole thing. "The dingy-dongy city people will bid the price up too high for us land-poor, millionaire farmers to afford," he grumbled. Sure enough, that happened. Garden farmers, looking for a place with a little land to raise their families, or beetle-browed executives, looking for a place to show off their wealth, could afford to bid $5,000 to $10,000 or even more per acre because they wanted only a few acres. The commercial farmer, wanting the whole thing, did not dare to bid much past $2,500 an acre because corn and soybeans weren't profitable enough to pay for it. So Marvin fussed. He conveniently forgot that in former years, when he was amassing his three thousand acres, he could outbid less-well-heeled farmers and city people alike. In those days he said that his ability to bid up the price on a farm was just "the good ole American way." Now that economics no longer favored him, it was no longer "the good ole American way" but a "bunch of greedy, sonsabitchin' land auction companies trying to squeeze every cent out of a farm that they could get."

George and Mabel paid $5,400 an acre for their five while Marvin fumed. He was not able to buy any portion of the farm. "There goes more land outta corn and beans," he growled under his breath. "Them dingy-dongy idiots are all gonna starve someday."

George and Mabel knew exactly what they were going to do with their five acres. They had planned their move for years while they saved their money. They had been reading about managed rotational grazing, and although they really wished they could afford at least twenty acres to practice it on, they believed they could do it on five by making every square foot of their land productive. First they fenced off a quarter acre for their house along the road. Then they put a fence around the rest of the property. Their plot of land measured 544 feet deep by 400 feet wide and, give or take a little for around the house, that meant about 1,800 feet of perimeter fencing. For that they chose permanent, woven wire fence. Marvin, passing on the road, scowling at what he called "white fence lawn maniacs," was surprised. Why were the dingy dongies putting up a good farm fence around their place? It wasn't long before the tractor farmer was cruising past the Cottagers' place every day, curious as to what was going on.

George and Mabel next built a little barn right in the center of the property that again mystified Marvin. Then they funneled the rainwater that would wash off the roof into two big watering tanks, one on either side of the barn. Marvin almost went into the ditch rubbernecking at the hard-working couple and their children. The Cottagers divided the rest of their land into four more or less equal plots, each a little more than an acre in size. All four accessed directly to the centrally located barn, and each water tank served two of the plots.

The Cottagers decided to go to the extra expense of using cattle panels for the interior fence instead of electric fencing. They felt that the cost of a little over $1,000 for the heavy wire panels was justified, because as fencing they would last at least as long as the Cottagers and could be easily moved if they needed to change the number and size of the paddocks. And they didn't have to worry about electric fences shorting out or shocking the neighbor children. Marvin noted the lavish use of cattle panels and muttered as he drove by: "A fool and his money are soon parted."

When the Cottagers planted fruit and nut trees around the fence line and sowed one of the fenced-in plots to improved bluegrass and white clover, Marvin smirked. "All lawn and trees on land that oughtta be growing corn and beans." When he saw the Cottagers lay out another of the plots in garden vegetables and corn, he frowned. "Must be some of those pinko commie organic nuts." But when the Cottagers broadcast oats, clover seed, and timothy on the other two plots, using only a garden tiller lightly over the plots before and after seeding, Marvin was totally mystified. "What the hell is going on here?" he muttered into the steering wheel.

Soon a horse appeared on the acre of grass. "Coulda guessed that," Marvin said, sneering. "Urban horse nuts." He was surprised when a calf, a flock of chickens, and a pig joined the horse. But then he thought he had it figured out. "Probably a petting zoo. Be a hundred school kids swarmin' over the place every week. At a dollar a head. Hmmm. Why didn't I think of that?"

He watched through the summer. The Cottagers would move their animals from one plot to the next as the forage was eaten down and then back to the first one to start the rotation over again. They made hay from one plot in June, mowing with a little sickle bar mower attachment on their garden tiller. They used leaf rakes to windrow the hay and then forked it into their pickup and hauled it into the barn. Marvin was amazed.

It took him until August to figure out what the Cottagers were doing. He had to admit that it was quite a trick. Especially tricky was how after one of the plots of oats went to seed, the animals grazed it and the clover to the ground. As the clover came back up, oat seeds that had been knocked to the ground by the grazing animals sprouted and grew another crop for oats, too. "By dingy dongy, did they do that on purpose?" he asked himself out loud. "Might be these people know a thing or two."

But it took him until December to get up enough nerve to stop and pull in the driveway. He had spotted something really weird. Parts of the garden plot had grown up in some very strange looking weeds, especially where they had dug the potatoes earlier and among the cornstalks, which the stupid organic nuts had not cut and fed to the animals. The thing about the weeds was that they were very green in *December.*

George was on his way to the barn. "Okay, okay, I give up," Marvin said, grinning ruefully. "Just what in the Sam Hill is that stuff out there?"

George laughed. "Turnips and kale. Good winter grazing for the animals. I was just about to turn them into there. C'mon, I'll show you." He led Marvin to the barn, opened the door that allowed access to the garden plot, and, to Marvin's amazement, the animals took to the green growth with great enthusiasm and then went into the dry brown standing sweet corn, gobbling up fodder and corn.

"Well, I'll be dingy dongied," Marvin said.

"I kind of bet your great-grandfather grazed turnips in the winter in the old days," George said. "I know mine did."

Marvin just kept staring at the steer tossing down turnips and gobbling corn fodder, and the hog, whapping down stalks of corn with a sideswipe of its nose to get the ears. In December. His mind was racing. Two thousand acres of corn and turnips. Hmmmm. Maybe feed out a couple thousand head of cattle over winter without so much as running one gallon of gas through the corn combine. Hmmmm. . . . and then out loud: "What you figure it would cost to fence a couple thousand acres?"

George laughed. "I'd have to get out a pencil."

"I think I know what you are doing," Marvin said, changing the subject slightly. He already knew about how much it would cost, and he wasn't about to do it. "Next year, you'll move the garden to the plot where your animals have eaten the clover down to nothing, right?"

George nodded, appreciating his neighbor's genuine interest.

"And the one this year in garden you'll rototill a bit and plant to oats and red clover next spring."

Again George nodded.

"And then the garden will move to the next plot the year after that, and then to the fourth, and start over again."

"Not to the fourth," George said. "That's permanent bluegrass, clover, and fescue. Will rotate just the other three plots."

"Fescue? Sounds like a disease."

"It's a grass that'll stay kind of green and grazable all winter. I'm trying to work out a system where the animals feed themselves nearly year-round. My motto is: 'Don't do anything that you can get a cow to do for you.'"

Marvin chuckled. He thought he might get to like this pinko commie organic nut after all. "I was wonderin'," he said. "That's a mighty fine steer you got there. Gonna butcher it yourself?"

"Yep. Actually more meat than we need. Know anybody who would want to buy some of it?"

"As a matter of fact, I do. And, if you need any help with the butcherin', I can handle that, too."

5

Good Fences Still Make Good Neighbors

The first step in establishing a pasture farm is putting a good perimeter fence around the entire grazing area. Don't even think about anything else until this job is completed. Although that should be obvious, its crucial importance is not well appreciated by beginners, especially those who want to maintain a small grass farm in an exurban countryside dotted with neighboring residences. There is no wrath like that of a serious gardener when your cows are thundering through her garden. As the population of the countryside in much of the United States becomes more and more an intermingling of exurbanites and farmers, good fences are a condition necessary for this new rural society to exist amicably. Even if it is no more than for the purpose of keeping dogs on their owners' properties, good fences still go hand in hand with good neighbors no matter what deeper thoughts Robert Frost was suggesting. Graziers farther out on open country may get away with only a fragile, single-wire electric perimeter fence, but I'm not going to promote that practice. I might run into the wrathful gardener at a book signing.

A close friend of mine, who believes the farm future belongs to monster tractors and ten-thousand-acre baseball diamonds to run them on, had a barnyard that looks very much like a farm equipment dealer's used machinery lot. That is his joke, not mine. He decided that a building to house his monstrously sized tractors, planters, grain harvester and headers, moldboard plow, chisel plow, disk, clodbuster, harrows, trucks, fuel tanks, spray rigs, fertilizer storage, and grain transporters was worthwhile even if it did cost $35,000. That's the least amount of money he could spend for a building big enough to shelter all that heavy metal. It is big enough to play a fast game of touch football inside it. And in many cases, as he admits, the machinery will be worn out or outmoded years before rust would ruin it from standing outside. Just looks bad, he says, having all that stuff out in the weather. And maybe, he muses, he can turn the building into a sports arena if tractor farming gets really tough.

I did some rapid mental calculations. A square 640-acre section of land takes four miles of fence to surround it. If you slice the section in two, each half equals 320 acres, about what my friend owns (the rest of his farming is on rented land). The rectangular half requires three miles of fencing to surround it. (There's a lesson here. One would think that half the acreage of a section would take half the fencing around it. Not so, when the half is a rectangle. To surround the greatest amount of land with the least amount of fence, use squares whenever you can.) I figure it is about 15,000 feet around the land my friend owns. With good woven wire at 34¢ a foot and good posts at another 26¢ a foot (spaced about 15 feet apart), a fence to go around his owned acres would cost about $7,500, less than a fourth of what that building costs. The fence would last about twenty-five years, costing roughly $300 a year. Cheap, single-wire electric fence to divide the farm into paddocks inside the perimeter fence would not cost as much as he now pays yearly for gas, diesel fuel, oil, and grease.

What's the point? For the cost of a building that will no doubt have to be replaced (at least the roof) about as often as the fence, he could start grass farming and get rid of his heavy metal disease. Yet whenever I argue for grass farming with farmers like my friend, they pretend that the cost of fencing is too high. The truth, if I use myself when I was a young man as an example, is that most farmers, especially those of us who grew up at the end of the horse era, have a cultural bias against fences. Fences are symbols of the past. With the advent of tractor farming, fences became an exasperation. Cattle and hogs were always breaking through them because the ones in place had been built by Grandpa thirty years earlier, and we didn't like to repair fences, let alone build new ones. Fences were always getting in the way. Trees and brush grew up in fence lines where sheep and cows no longer grazed them. Trees and brush sapped moisture from nearby crop plants. We ran our super-duper machines into the fence while turning (half asleep from driving too long) at field ends. Unaccustomed to how wide the new cultivating tools were behind our super-duper tractors, we regularly snagged them in the fences and tore out great gaping sections or damaged the machinery. I don't know how many field gates I cursed in earlier days because they were too narrow for our latest, big, whoop-de-do, grain harvesters. We had to widen all the gates on the farm. In some instances, we just knocked the gateposts over. To hell with it. We weren't going to pasture animals anymore anyway. We would pen them up in modern "dry" (ha, ha) lots and bring the feed to them. We were the Future Farmers of America. (In the new millennium, FFA has come to stand for "Fathers Farm Alone.") There was a giddy sense of freedom in being able to look across the land and not see one damn fence all the way to Indiana. Our chains of slavery were dropping away. Expand. Away with fences. Our favorite song of the nineteen forties was "Don't Fence Me In." And although the songwriters didn't realize it, expansionist farmers were the reason that song stayed at the top of the Hit Parade for so long.

Woven Wire for Perimeter Fencing

Most graziers today will probably opt for one of the substantial, five-wire (or more) high-tensile, electric fences for their perimeters, a decision I think they will regret. A New Zealand–type fence, as this is usually called, is a *little* cheaper and somewhat easier to install, but it will never be as effective as a stout woven wire stock fence. Inevitably, calves and lambs and especially goats will "leak" (the term graziers use, and they don't mean urinate) through an electric fence no matter what. Manufacturer claims to the contrary, electric fences will short out if enough weeds grow up under them. And enough weeds will grow up under them, especially on the side away from the grazing animals. The grazier will spend considerable time and money either spraying these weeds or mowing them. In the long run, the cost will be more than a good, impregnable woven wire fence. Yes, weeds will grow up in woven wire, too, but there is no electric charge to short out or drain your charger.

Permanent fences are a pain in the neck no matter how you manage them. The weeds and brush do grow up in them faster than you can cut or spray them out. Deliberately growing trees along a perimeter fence, like I do, is some kind of insanity. Tree limbs break off in storms and short out electric fences. Whole trees blow over and break down even woven wire fence. Some trees are worse than others. Weeping willow and black walnut are two species definitely not for planting along perimeter fencing. The wood is brackish and limbs of even younger trees break off in storms. I would never give up the cedar hedge windbreak along my western perimeter fence, but I bow to the necessity of cleaning up branches that sag, heavy from ice and snow, and bend down the woven wire fence. The cedars would be certain death to an electric fence.

Cows and especially sheep take care of weeds in fencerows to some degree, but not so well under perimeter fences unless the neighbor on the other side also grazes livestock. Spraying brush killer at the base of perimeter fences every three years is the easiest way to keep the fence line clean. I use lawn mowers and weed whips, which along with the sheep nibbling around the base of the fence, suffices most of the time.

As I said above, woven wire is a little more expensive than high-tensile electric fence, at least initially, and it is harder to build correctly, but it is the sturdiest fence for the money (chain link is of course sturdier but too expensive to justify in most pasture farming situations). With a single electrified wire above it, woven wire is practically impregnable to farm animals. A good woven wire fence discourages coyotes and dogs, too. It will stop snowmobile and four-wheeler idiots who consider fences an affront to their freedom. It will not stop coon hunters, however, who have a habit of cutting holes in woven wire fence to let their dogs through. Coon hunters have their own law about private property. "If it's dark, we own it." I don't mind, really, because raccoons are the single most destructive animal on our farm, and now that coon hunters are almost a lost tribe, I sometimes wonder if the coons will not take us out someday.

A woven wire perimeter fence is the key to the grazier's tranquillity. The consequences of animals getting loose on the farm are not so bad in pasture farming, but animals loose on a highway or a neighbor's garden can mean big trouble.

I say all this knowing full well that out in the country, where there are few if any residences or highways nearby, graziers are using fragile single-strand electric polycord or the wider, ribbonlike electric cord for their perimeter fencing as well as for their cross-fencing. Cows brought up on electric fence and moved to new lush pasture frequently can be secured this way. They become very aware of the fence and will even kneel down and stretch their heads under the single strand to get a bite of clover in the next paddock, but even at the farthest reach of their necks, they will not touch the fence. It is very comical to watch them do this. Because they graze the pasture directly under the fence and, in fact, a foot or two on the other side by reaching under in this way, weeds rarely become a problem, especially if the fence is moved over occasionally and the forage mowed where it regularly stands.

For the pros, all well and good. Garden farmers who are working away from the farm during the day, or who must take business trips that keep them on the road overnight, would be wise to put a substantial woven wire fence (or a fence of cattle panels, as I will explain below) around the borders of their land.

The difficulty with woven wire is that putting it up calls for methods that are not readily known anymore, whereas there is a ton of information available on electric fencing. Woven wire fencing is made of crosshatched wires with mesh openings from 6-by-12 inches to 6-by-6 inches with variations. For a perimeter farm fence, install what the catalogs call cattle fencing, about forty-seven inches tall (depending on whether you locate the bottom of the fence on the soil level or an inch or two above it, as I prefer), with the horizontal wires going from three inches apart at the bottom to six at the top. This bottom spacing is crucial. Do not, as I did at first, try to save money by buying fencing where the space between the horizontal wires is the same six inches at bottom as at top. Sheep and goats will shove their heads through a six-inch spacing near the bottom to get at grass on the other side and eventually loosen the fence and/or get their heads caught. Even if you do not plan to raise sheep or goats, buy the proper fence anyway, because you don't know what you are going to raise in the future. The difference in cost is not that much.

Mesh with vertical stays six inches apart is best, but eight or twelve inches is okay. Sometimes sheep will get their heads caught in six-inch square mesh, so keep a sharp eye out or choose a wider mesh between stays. Sheep are very stupid about fences. They will thrust their heads forward through them with considerable force, but they will not always back out with considerable force. I have used a lot of six-inch mesh fencing because I got it free when the highway department put in new fence along the interstates. Since this highway fencing is made with all nine-gauge wire, thicker and stiffer than the usual woven wire with eleven-gauge middle wires, the sheep are not as prone to push through it. Of course, a good grazier would say that if my sheep are reaching for grass on the other side of the fence, it means I should have moved them to a fresh paddock sooner. Maybe so, but I think sheep are like humans in this respect. The grass is always greener on the other side—even when it isn't.

The gauge, or thickness of the wires, is the third crucial decision in buying fence. The top and bottom horizontal wires should be nine gauge. Anything skinnier won't last long enough to justify the labor of putting up the fence. The middle wires can be eleven gauge (the lower the number the thicker the gauge). All nine gauge is much better but more expensive.

The amount of copper in fence metal is also important to its longevity. My experience is that hardware and farm supply stores have no idea what copper content is all about anymore and the information is not easily available at retail stores. You can't tell by looking, either. I just buy the best quality Red Brand or similar fencing. If there are two rolls of fence for sale that appear to be exactly identical in every way, but one costs $100 for a twenty-rod roll and the other costs $120, buy the more expensive if you are thinking in the long term. The kind of woven wire the highway departments use (all wires nine gauge) will last forty years as a pasture fence, but I think the cost would be prohibitive for fencing large areas. (I don't even know where you can buy it.) What I've gotten for free was twenty years old when the highway replaced it, and it is going on fifteen years on my place. Another way to save money on fencing is to go to farm sales and buy the used rolls and posts often auctioned off. Sometimes there are new rolls offered. In areas where there are not yet many pasture farms, the fencing will go cheap. But make sure it is not rusted too badly. Putting up rusted fence is a lot like putting new wine in old wine sacks.

The metal posts that the highway uses cost upward of eight dollars each, too expensive for graziers even though they will last long enough to make up for the higher price. They are much heavier and more rust resistant than normal posts. Some I have are going into their forty-fifth year with little sign of deterioration, even at the soil surface, where posts normally rust off.

A good metal post heavy enough to last twenty years costs about four dollars now. Cheaper ones are not worth the work of installing them. You want the tallest ones, seven and a half feet in length. Drive them into the ground two and a half to three feet, leaving enough above ground for a forty-seven-inch-tall fence and a strand of barbed wire or electrified wire above it if you have horses or cows. There are plastic posts available that won't rust, but they are not strong enough or long enough to suit me.

I prefer wooden posts, but digging the holes with a posthole digger or auger is slower, harder work than driving metal ones with a metal post driver. In rocky ground, digging holes for wood posts can be difficult to almost impossible, but you can rent hydraulic power drivers to ram sharpened wood posts into the ground. Very nice but more cost. Wooden posts at lumberyards are usually made of cedar treated with a preservative. In some areas, black locust posts are available and are usually a little cheaper than good metal posts. Hardly ever available are catalpa or Osage orange, although they are as long lasting as black locust. Black locust and catalpa will both last forty years, but the latter is better because it is softer wood, making it easier to hammer staples into. A hundred years ago, nearly every farm in our neighborhood had a catalpa orchard, specifically for fence posts. (It is great wood for carving, too.) I have posts that my grandfather split for his fence, my uncle used a second time, and I, a third time. A few still in use are more than sixty years old.

Currently, we are fencing my son's new farm. To cut the cost of woven wire fencing by half, we have been able to find discarded electric poles and railroad ties for free. The electric poles, especially the butt ends, which are heavily creosoted, make marvelously long-lasting fence posts. We can usually split out six posts from one eight-foot length of pole. Believe it or not, railroad ties generally split into two posts rather nicely. Splitting out posts is hard work, but I can be greatly motivated when I know that one length of free electric pole will make six fence posts better than the ones you buy for $3.99 each. In something like thirty minutes, with time-outs to catch my breath, I can make $24.00. Young men can double that output. We use a tractor-powered auger to dig the holes, and when we really get going we can put up a lot of fence in a weekend.

Corner posts and end posts for woven wire should be at least eight inches in diameter and nine feet long and made of treated wood or wood known to be long lasting, like those mentioned above. I've occasionally used black walnut that was not good enough for furniture because it lasts a long time in the ground, too. The end posts need to be sunk in the ground at least four feet. Nine-foot lengths of electric utility poles are ideal. Corner posts at the lumber yard aren't as thick and cost up to twenty dollars last time I checked.

Corner posts and end posts need not only to be put in the ground four feet or more, but also to be braced mightily because the pull on them from a woven wire fence, stretched as tightly as it needs to be, is terrific. We put a second post about ten feet out from the corner or end post, in line with the fence, of the same size as the corner post, sunk to the same depth. A brace is notched to fit horizontally between these two posts toward the top of each, and a double strand of nine-gauge wire or thicker is strung around the bottom of the end or corner post and around the brace post at its upper end. Then, with a stout stick, we wind the strands together until they pull both posts up tightly to the horizontal brace. Do not try to use those decorative rails that make such pretty fences around suburban properties. They will warp and rot. The rule of thumb for woven wire fencing says the length of horizontal braces between end post and brace post should be two and a half times the height of the fence.

With the end or corner posts in place, and the dirt tamped solidly around them, we proceed to put in the wooden posts or metal posts between. I sometimes put posts as far apart as 18 feet, but 15 is better, and 12 doesn't hurt anything except your pocketbook. The fence must be perfectly straight between the posts to get a strong, permanent stretch. The slightest curve can cause the fence posts to lean out of kilter.

We roll out the fencing alongside the posts, attaching one end of the roll to the end post or corner post and rolling the fence out past the other end post or corner post. We stick a temporary post in the ground behind the end post, and the fence stretchers are attached to it and to the fencing to be stretched. It is easier to stretch the fence this way than to attach the stretchers to the end post itself. In the latter case, the stretchers get in the way of stapling the fence to the post after it is drawn tight. To splice two lengths of woven wire together, there are available little clamps easily attached to the wires to be joined. Or one can wrap the ends of the two wires around each other very tightly.

I hesitate to get into the minute details of stretching a fence because you are going to just have to do it to learn them. If you can get help from someone who has done it before, so much the better. My next-door neighbor raises livestock, too, and we had a mutual interest in putting up the fence between us. And he knew how. Incidentally, the old fence laws that regulated the responsibilities of each landowner for a line fence are no longer enforced because tractor farmers or neighbors without livestock usually refuse to maintain their half of a property line fence. Where both property owners have livestock, no law is necessary. They will willingly share the cost of the fence.

Fence stretchers for woven wire are available from farm supply outlets like Nasco Farm and Ranch catalog in Fort Atkinson, Wisconsin, or Modesto, California. Manuals and books of instruction can also be purchased or come with the product. You will need them unless you have helped install a fence before. I was able to purchase all my fencing tools either at farm sales or privately from retiring farmers.

As you ratchet up the stretcher(s), the fence gradually tightens. We generally raise up the fence manually to a half vertical position after the tightening has progressed to the point where the fence is tilting upward of its own accord. Then, with a piece of wire about every tenth post, the fence is held in that position. More tensioning and the fence will stand vertically on its own, and you can release those holding wires. Knowing how tight to stretch the fence is something only experience can teach. The horizontal wires of a woven wire fence, especially the top horizontal strand, are slightly kinked. When about half the kink has straightened out in the stretching, the fence is usually tight enough. When the end posts, in the ground four feet or more, start to give a little at ground level under the tension, that also means the fence is tight enough. The fence by then will be very stiff and rigid. If you are stretching over a rise or fall in the ground surface, you will have to stop stretching when you can still lift up or push down the fence to attach it to the post. You can't stretch up a hill and down the other side in one section but must stretch from the bottom to the top, then stretch another length of fence from the top back down to the bottom on the other side of the hill. This may sound a little mystifying, but out in the field, on the job, you will very quickly understand what I'm saying. You must stretch in relatively straight lines, not around curves or over hilltops.

There are special wire clips to fasten fencing to metal posts. Use common two-inch staples to nail the fence to wood posts. Wrap the wire completely around the end posts and attach the wire ends back to the fence. Then staple. Always staple the horizontal wires, not the vertical ones. Except for the end and brace posts, don't drive the staples solidly into the wood. Leave a wee bit of space between the staple and the wire for the fence to move. Metal expands and contracts in cold and hot weather and the staples loosen if they are driven in solidly. The staple should be nailed so that the two prongs bite into different grain in the wood—a little sideways rather than straight up and down.

For animals the size of horses and cows, you will need to put either a line of barbed wire or electric wire about six inches above the woven wire to keep these animals from reaching over the fence and eventually riding it down. Electric is more effective. On top of the woven wire, or along the inside of the fence, the electrified wire is somewhat out of the way of children. A hungry horse will ride down a barbed wire, but even after a woven wire fence gets old and rusty, the electric wire on top will keep large animals at a respectful distance.

Electric Fence Details

As I say, many pasture farmers, especially those not near populated areas, will no doubt choose some form of high-tensile electric fences or single-strand polycord for their perimeters, because over several miles the savings compared to woven wire add up. Remember that with electric fence you have the cost of a fence charger or energizer to figure in. There is plenty of information on installation available for electric fencing. Some graziers use five single wires or more about ten inches apart, starting at six inches above the ground for sheep, and electrify several of them. For cattle, the bottom wire should be about a foot above the ground and the top one about eye level to the animal. Where you save money is in the posts. They can be as much as one hundred feet apart.

There are also many tools and gadgets now to make the job of putting up the high-tensile fence easier yet. Special stretchers, or tensioners, tighten the high-tensile wire with hardly more than a flick of the wrist. The pace of change is so fast in electric fence construction that much of what I could say might be outmoded before this book gets around, so I will not waste your time here. The advances in electrified fencing, in fact, are what is making commercial pasture farming practical and profitable. Improvements in solar-powered batteries to supply electricity far from utility outlets are particularly significant. Some of the new fencing looks to me to be too fragile, but that remains to be seen. Use your head. Talk to other graziers. If you are new to fencing, refer to advertisements in grass-farming magazines like the *Stockman Grass Farmer* and follow up on the latest information they supply.

There are many different kinds of insulators to allow attachment to posts without grounding out the wire. Sometimes plastic posts are insulative themselves and don't need insulators. There are all kinds of wire, including electric netting and stuff that looks like tape (looks ugly to me), all aimed at reducing the cost of fencing without losing effectiveness. For movable electric fencing in strip-grazing, graziers generally prefer polycord, which looks and handles more like tough cord than wire and is very light and very easy to loosen and reattach. For strip-grazing, a field is generally divided into long rectangular sections, with a single electrified wire and small, slender posts down both long sides of the paddock. Then the rectangular plot is grazed lengthwise, by moving a single polycord wire that stretches between the wires that form the sides of the rectangular paddocks. Every day, the polywire, or polycord, is moved forward into new grass the distance necessary to give the cows a good daily ration. The cows become accustomed to this arrangement, going straight to the new grass available to them when the cross-wire is moved. They will even stand by the wire, waiting patiently for their master to move it. There are new ideas constantly being applied. One kind of fence actually rolls along on tall wheels to make moving the wire every day faster and easier. When one rectangle has been grazed from end to end, it is closed off and the rectangle next to it is strip-grazed. By the time the last rectangle has been gone over, the first is ready for another round of grazing.

Some electric chargers or energizers are run by DC batteries, including solar-powered ones; some use regular AC current. This is another case in which cheaper models can end up costing more money. Voltage is not the crucial number to look for; instead, look for the joule rating, which is imprinted on the energizer. If you don't remember your high school physics, you'll have to look up joule in the dictionary like I did, but you don't really need to. The rule of thumb is one joule per six miles of fence. Also important is the length of the pulse of the energizer. Electricity does not flow steadily through the fence but in spurts. A good energizer has very short pulse lengths: .0003 second. A cheap energizer has longer pulses, as much as .3, which is not good. The longer pulse can start fires and melt polywire that is shorted out. Both AC and DC battery power work fine, but DC batteries are a whole lot more practical out in a field far from an AC outlet. On the other hand, DC batteries out in a field far from human habitat are great temptations for vandals.

Heavy Wire Cattle and Hog Panels for Fencing

We are finding that hog panels, or cattle panels, are most useful for special fencing situations (and maybe eventually for all situations on small farms). These panels are made of crosshatched metal of about pencil thickness, the vertical and horizontal wires welded where they cross. The metal is galvanized or specially treated to resist rust. The wires will last forty to fifty years in a fence, maybe longer. They are usually sold in sixteen-foot lengths and cost about a dollar a foot. Sometimes they are on sale for less. They are stiff enough to stand alone but are light and flimsy enough to bend. I know an orchardist who bends one sixteen-footer around each of his new apple trees to discourage deer. The ones for cattle are about 48 inches tall; the ones for hogs are about 36 inches tall. The mesh is 6 by 6 inches, but on some kinds the horizontal wires are closer together at the bottom to keep little pigs and lambs from squeezing through. In a fence, they work well with a post every eight feet and the panels overlapping two inches or so at the ends. I suppose you could cheat a little and put the posts ten feet apart.

As field gates, cattle panels are cheaper and lighter than wooden board gates and easier to open and close. I have found them useful for putting around haystacks to keep the animals away when you don't want them eating the stack or to control how they eat the stack without wasting hay. The panels also make easy-to-install floodgates across creeks because they are light but strong. Sometimes one sixteen-footer is enough, sometimes two. I drive metal posts at the bank of the creek and wire the panel to it on one end. The other end, out in the water, I may wire very loosely to another stake or not at all. When high water rolls through, it pushes the panel open to allow debris to flow through. When the water goes down, all I have to do is close the panel again. I can easily remove the panels in winter freeze-up and put them back in place when spring comes. Sometimes we cut the panels in two so the pieces fit better up the creek banks.

We have even started using the panels for regular fencing in special situations, even though the cost is twice that of woven wire. Constructing the fence is simple. Just set the posts and then wire or staple the panels to them. No stretching. Since the panels last so long, are easily moved to new locations, and allow entry into a field wherever one wishes, the extra cost really turns into savings over any other kind of fence in a span of fifty years.

We figure we can at least afford the extra expense where we are fencing through or alongside woodlots or over rough terrain where it is almost impossible to stretch a woven wire fence. We can carry the panels easily into places where we could not even unroll woven wire. We can weave in and around trees or big rocks because there is no need for a perfectly straight fence as when stretching woven wire. We have already experienced an unforeseen benefit of great significance. A tree fell over the fence recently and crushed one panel. Had that been a woven wire fence, extensive repair would have been necessary, and the fence never would have stood as strong and stiff again. But with the panel, all we had to do was cut away the fallen tree trunk, bend the crushed panel back to its proper shape, and reinstall it. The repair took hardly a half hour.

If I were starting a very small grass farm, say of five acres, I am quite sure that I would do all the fencing with the kind of cattle panels that have shorter spaces between the horizontal wires close to the bottom, called combination panels. I would then have to put up a perimeter fence once in my lifetime, and I could move cross-fencing with ease. I am absolutely sure that as the woven wire fence in my cedar tree fence line deteriorates, I will replace it with cattle panels. I'll just wire them to the cedar tree trunks. I would not have to worry about livestock or even hogs breaking through it since the panel wires are much stronger than the best woven wire. Not even a horse could ride down a cattle panel by trying to reach across it. And with combination panels, horses are less likely to get their legs stuck in the mesh. Nor can large dogs and coyotes get through it and into the paddocks. Nothing would need to be electrified, so I would never have to worry about the fence shorting out. The only worry is that on rare occasion a sheep might get its head caught in the wire mesh. I would have an infinite choice in the size and shape of paddocks because the panels are so easily moved. I presume also that if I bought the panels in large quantities, I could get a cut in price.

The panels can be cut to smaller sizes with heavy wire cutters, a hand grinder, or a hacksaw. As four-by-four- or four-by-six-foot pens, the panels are handy for transporting sheep, goats, large dogs, and other animals of that general size in a pickup. You can easily make a temporary lambing pen out of one by bending it around in a circle. A larger pen can be made with two or more panels.

With a good perimeter fence in place, I daresay you will have accomplished three-fourths of the work of successful pasture farming and will have eliminated almost all of the worriment. You can go on a trip with a free mind. You can sleep well at night. I speak with some experience. Once I was grazing about twenty unbred heifers on my grandfather's farm, which, unfortunately for me, lay right on the edge of town. Unfortunately for me again, Grandpa was not the greatest maintainer of fences, and neither was I in those days. To make a long story short, there is no adventure quite like riding in the back of a police car, chasing a bunch of heifers down Main Street with a contingent of the village's population in hot pursuit. The Running of the Bulls in Spain had nothing on us. Heifers can run faster than bulls, for one thing. We surrounded the animals more or less in the parking lot of a fast-food restaurant. By the time the heifers had taken stock of the parking lot and realized that they were surrounded by cars, they decided they had had enough fun and were content to follow me, or rather the bucket of grain in my hand, out of town and back to the farm (giving new meaning to that term). Meanwhile, before and behind me, like a parade, police cars eased along with red lights blinking. That is how I learned that good neighbors still make good fences, which is what Robert Frost really meant.

6

Water in Every Paddock

After fencing is in place, at least perimeter fencing, the next important step is providing good water for your livestock. The fastest way to a summer slump in weight gain or milk production is by not having fresh, clean water nearby for your animals. Ideally, water should not be more than about one thousand feet away at all times, and it is better if the distance is around eight hundred feet, so say the experts. After water has done its job of slaking thirst, it becomes a very significant fertilizer in the form of urine.

In a very small "backyard" pasture garden, supplying water to several paddocks is easy and inexpensive: a hose attached to a spigot will do the job. Or you can use a galvanized water trough or two to catch water from the roof of a little barn placed strategically at the center of the paddocks, as described in chapter 3.

On larger farms, supplying water to every paddock is more involved. Generally, graziers pump water to tanks located under the single polycord electric fence that separates pastures so that each tank or trough serves two or more paddocks. Often, it takes a number of years to learn just what design and layout of a water supply system work best on a particular farm. There is no fast way to accomplish this, nor is there any written instruction that can take the place of experience. I have not yet completed my water system after five years of experimentation, but I now know what I should do. I just need the money with which to do it. In the meantime, because I have such a small farm, I can easily carry or haul water to the paddocks not yet supplied with their own water.

Whether you pump water to tanks, make use of springs, use windmills, or dig ponds depends on your situation, of course. In the north, you will have to provide for some way to prevent ice from forming over the entire pond or tank, although spring-fed ponds might not freeze. Water heaters can be supplied for tanks in some cases. Ranchers sometimes use wind-driven agitators they call pond mills to accomplish this task. I just chop a hole in the ice when necessary. You may have to do this on creeks, too, but ours is fortunately fed by springs, and only in the very coldest weather does it freeze over, and then not for long. Freezing is not as big a problem as it might be, because sheep and beef cattle can get by (not more than ten days, most graziers say) eating snow in an emergency. Even in summer, when the grass is dewy, grazing animals will drink less.

Because of the growing interest in pasture farming, manufacturers are supplying simple and rather inexpensive water systems. For example, high density plastic pipe is easy to fit together and reasonable in cost. It does not normally burst if water freezes in it. It comes with longer-lasting valves and easy-to-attach couplers. Once you start shopping for the equipment you need, you will find tons of information about how to proceed. As I have said often in this book, the fastest track is through the advertisements in the *Stockman Grass Farmer* or other grass farming publications.

Creeks as a source of water may work well for many graziers, but *caution* is the word. There is much controversy over the use of creeks by cows because a large herd can pollute the stream. Unlike sheep, cows can get into the habit of standing in the water during the hot part of the day. The controversy comes mainly when zealous conservationists try to apply the rules made for ever-running creeks to those that dry up in the summer. It doesn't make sense to a grazier to talk about stream pollution if a stream is dry part of the year, usually the part when cattle are tempted to stand in the water. Kansas (and perhaps other states) has made laws contradicting EPA pollution regulations for cattle in streams when the flow is intermittent. In other situations, many ranchers and graziers of their own volition figure out ways to use streams without polluting the water. One new way to do that is with solar-powered pumps that can be hauled to a stream far from an electric outlet and used to pump water into a creekside tank. At least one manufacturer offers a pasture pump that activates itself, moving water, as much as 26 feet vertically and 126 feet horizontally, so the advertisements claim (Rife Manufacturing, P.O. Box 70, Wilkes-Barre, PA 18703).

All things considered, I believe that ponds are better than creeks for watering animals. The trick is to build ponds or use natural ones, so that each serves more than one paddock. Years ago I often observed dairy farms in Minnesota, the land of ten thousand lakes, that would utilize the large lakes that bordered their properties for stock water. Any number of pastures might open to the lake, with fences running far enough out into the water so that the cows would have to swim to get around them, which they never did. Most of the lakes I knew then (and where I would go skinny-dipping with the cows) are now ringed by houses. Those beautiful, bucolic farms, those beautiful bucolic days, all gone. All gone. What a pity.

Another possibility not commonly pursued yet (because pumping water to paddocks seems cheaper) is the installation of several small ponds. The terrain on many grass farms in humid regions would allow for many small ponds dotted over the pastures. Such ponds would need to be only about fifty feet in diameter, like ours, and cost no more than five hundred dollars each to install—what ours cost. Where rainfall averages thirty to forty inches a year, such small ponds would require only a watershed of a couple of acres to supply enough runoff water to fill the pond. Soil from the excavation could be used to funnel runoff water into the little pond or away from it. I am about to put in a second pond to prove it.

Ponds as Fish Pastures

There is another way to look at ponds used primarily for watering livestock. A pond is, itself, literally an aquatic pasture, a paddock where one can graze fish, frogs, turtles, ducks, and other water animals.

At a conference recently, I asked the rural audience what part of their farming they enjoyed most. By a show of hands, the majority said their farm ponds. (Permanent pastures were the second favorite!) The pasture pond, almost as much as the kitchen, was the center of home life. A pond provided swimming, fishing, ice skating, picnicking, nature watching, and a place for family gatherings.

Mind you, I do not have in mind a commercial fish farm, although that is always a possibility, and one I believe could mesh well with grass farming. As with my pasture farming, I want my pond farming to be a laid-back enterprise to produce food for several families, not a profit-making enterprise. I don't want to go overboard (oh, those sneaky puns) on fish stocking rates in the pond any more than on animal stocking rates in the pasture; but I do want the pond(s) to have agricultural purpose beyond recreation. As a watery pasture for a flock of fish (whoever coined the phrase "school of fish" never tried to teach a bass to read), the pond appears to me to be the most productive "paddock" on the farm for the amount of space, work, and expense involved. As in grazed pastures, the "livestock" do most of the work. The grazier's "work" is fishing.

For warm-water ponds like ours, largemouth bass and bluegills have proved to be the best livestock. If I had a spring-fed, cold-water pond, I would try trout. I think yellow perch taste better than trout, and they will thrive in some warm-water ponds, but for some reason we have not discovered yet, not in ours. To control weeds, I stocked a couple of grass carp (white Amur). They eat almost all vegetative growth in ponds, including cattails. One grass carp is enough in a pond as small as ours. Grass carp might be the missing link between fish and livestock if one were to believe Chinese folklore. Chinese aquaculturists like to say, with a little smile, that grass carp will eat all the plants in the water and then flop up on the bank and graze the grass awhile before flopping back in the pond. Needless to say, I have not seen a grass carp do this, but when I circle the pond with the mower, clipping off edge-water grass, the carp circle with me, eating any grass clippings that fall into the water. In Asian countries, the fish are fed grass, and once they are fat they are caught, butchered, and sold like any other livestock.

The only thing that grass carp won't eat are the yellow pond lilies I introduced because I didn't know any better. Do not do this. The lily pads make wonderful shade for the fish and provide oxygen, but I have to keep cutting them back (saw them off with a corn knife as far down in the water as I can reach without drowning) to keep them from covering most of the little pond. The pond lilies are not supposed to take over deep water (eight feet or more), but I know for sure they will grow in five feet, and that's most of my pond. Actually, cutting back the lilies is not unpleasant work, but what I should have done is grow a dozen lilies in pots, set the pots in the water over summer, and then remove them. Of course, if the blossoms spread seed into the water, plants will get established in the pond anyway. I suppose there is some poison to kill the lilies, but I fear I might kill the fish, too.

Almost all fish overpopulate in ponds except the triploid grass carp, which is genetically altered so that it can't reproduce. Channel catfish don't usually reproduce in ponds and so are a good introduction if you like the taste of catfish. Other catfish will overpopulate in a hurry. You can get plenty of information from your local Farm Service Agency (FSA) office on stocking rates for various fish, especially bass and bluegill, which are considered the best combination for warmwater ponds. But, because fish can't read, they rarely conform to the expert advice. The expert advice rarely takes into account the holistic nature of a pond but focuses only on the predatory relationship between bass and bluegill, as if these fish were living in a test tube and not a pond. In a natural pond there are many other predators besides man, and they can knock a hole in anybody's stocking rate calculations: turtles, ducks, raccoons, bullfrogs, snakes, herons, and kingfishers, to name a few, all help themselves to the fish. Largemouth bass not only eat their own fry but anything else they can stuff in their cavernous jaws. There are on record accounts of largemouths leaping out of the water to catch small birds. I once saw one of ours grab a male dragonfly clinging to its mating female as she hovered over the water. She seemed relieved as she flew away.

Overpopulation is not a problem for us, because we manage the pond for fish the same way we manage a paddock for livestock. We know by counting when there are more fish than we want in the pond. The water is clear enough to see them most of the year. We hand-feed worms to the fish, and they come swimming to us eagerly. That makes not only counting easy but also catching them on a hook. Initially, we stocked the pond with twenty largemouth bass about six inches long and half again as many bluegills and red-eared sunfish, plus two grass carp and six perch. All have survived except the perch. One of the grass carp we caught and ate along with routine catches of the bass. I take constant note of the young fish, which I call yearlings. Most of them get eaten, so we are not threatened with overpopulation.

Our fish do not grow fast. We could increase rate of gain by supplemental grain feeding, just as we could the land animals, but our fish won't eat commercial feeds from anyone's checkerboard square. Actually, I'm glad of that because it forces me to stick to my first intention: to raise fish in an entirely natural, self-subsistent way, the way I want to raise livestock, too. I asked around about why my fish are so contrary as to refuse supplemental commercial feed. One expert said it was because they had been around me too long. Most said that fish prefer insects and minnows and other naturally occurring food. They have to be forced or trained to eat grain feeds. Fish hatched in artificial ponds and started out on grain will eat it and thrive. I guess mine find enough bugs and minnows in the pond, because they will not even eat bread crumbs or, snooty things, sweet rolls. They will mouth bits of these offerings and then spit them out again. I held the book that says they like pastries right down close to the water for the fish to see, but, like trying to get a conservative to consider liberal writings, they would not read. School of fish, my eye.

Our "breeding flock" loves worms and crickets and grasshoppers. I use a butterfly net to catch the latter two out of the grass around the pond, and we try to find time to dig and feed a can of worms occasionally. The grass carp are supposed to be vegetarians, but evidently they can't read either, because we caught one of ours on a hook baited with a worm. We baked it. It did not taste as good as the Chinese claim it does. I have been told that I should have kept it live in a tank of clean water for a day or two before cooking and eating it. If you can find a good Chinese recipe for grass carp, you could overstock them in your pond, feed them grass and hay, and supply the whole neighborhood with fresh fish. They grow to prodigious size. If you wait fifteen years, you may have to lasso them with a hay rope and haul them out of the water with a front-end loader. Grass carp may also eat fish eggs, according to some sources. I place a couple of old Christmas trees in the water every spring so that there are places among the branches where the other fish can lay eggs and where the carp—and ducks and geese and herons—can't get to them. A few minnows survive every year to grow into yearlings, so I think the food chain is fairly well balanced.

We also covered the inner banks of the pond, down about three feet into the water, with rocks (called rip-rapping) to protect the banks from trampling livestock and erosion and to give the minnows places to hide from the big fish.

The remarkable thing about our tiny pond is that we have not had to supply supplementary oxygen even in the bad drought of 2002, when the water in the pond shrank by half, to only four feet deep. I can't explain why, except to theorize that the fish population has not risen above naturally sustainable levels. I believe that ponds need aerators (there are many kinds available) only when pond owners want to grow more fish than natural levels allow. Thinking about my fish, I also am thinking about my land animals. Trouble starts when the population level of either is pushed beyond natural limits.

Aerators definitely defend against fish kills caused from a "turnover" of the water in larger, warm-water ponds. Turnover is rare in ponds as small as ours because the depth in summer is not enough to maintain a zone of cold water on the bottom of the pond. In deeper ponds, when the colder bottom water, which is low in oxygen, rises suddenly to the top due to a quick change in the temperature of the surface water, the mix of water causes a lowering of overall oxygen levels and fish die by the hundreds. This high mortality almost always occurs because the fish population is too high. It is nature's way of getting rid of excess fish. A parallel phenomenon for getting rid of excess population in the human situation is ozone or other pollutants trapped near the earth's surface by barometric pressure.

Most water-quality problems can be avoided by positioning the pond where surface water running into it comes only from pasture and woodland and not from cultivated, erosion-prone, or chemicalized farm fields. This is why ponds are especially suited to pasture farming.

In some instances where a well is handy, you can design a pond so that runoff water can't flow in. Fill it by pumping in groundwater. I sited our pond only after several years of observing water runoff over our fields. In one pasture there was a slight depression in a very gently sloping field. Rainwater flowing over that spot came only from the meadow itself and a woodlot. The watershed was about six acres, more than enough for our small pond. We dug the pond in that depression, a simple hole in the ground, what engineers call a "dug pond" rather than an "embankment pond" that needs a dam. Clean runoff water coursing across the pasture during rains fills the pond and then oozes over the bank at the lower end only very slowly. There is no problem with washing away the bank, nor is there the expense of building a concrete outlet box and emergency spillway, which drive up the cost of dammed ponds. For building dammed ponds, information galore is available. See your local FSA office. Or talk to earthmoving contractors who have a good reputation for building ponds in your local area.

The general rule states that for a pond in our central Ohio climate, the water should be at least eight feet deep at the center for fish to survive winter. Although that is a good rule to follow, it isn't really true. In more than one drought year, we have gone into winter with hardly four or five feet of water, and in twelve years of varying weather I have found only two winter-killed fish.

I have learned something that I had not heard in all my years of living around farm ponds. The bass and bluegills and grass carp "disappear" as winter comes on. They reappear in spring. They burrow into the bottom mud for the winter in a state of semihibernation. I don't think they know or care how much ice is above them as long as there is enough water to keep some algae and tiny waterbugs alive. I do scrape snow off parts of the ice so that sunlight penetrates to encourage a little oxygen-producing algal growth even in winter. I have never seen ice freeze more than ten inches in our climate. When the ice gets that thick, it acts as its own layer of insulation over the water below it. The ground temperature of a more or less constant 55 degrees Fahrenheit keeps the ice from freezing deeper. I have read of northern areas where ice gets much thicker, but I find that hard to believe.

A watertight soil is a must for larger ponds, although sometimes a borderline soil that leaks a little at first will seal over naturally. It is not a good idea to bet on that happening, however, without expert advice. But for small ponds like ours, a sandy or gravelly soil need not be a drawback. The cost of a liner is not prohibitive these days when you take into account that it will last many years. A nephew of mine put a dug pond on a hillside where the subsoil was mostly shale and leaked like a sieve. The pond is twice the size of mine, and the liner cost him $1,500 including a layer of felt underneath to keep rocks from gouging holes in the liner. He figures that cost is reasonable as a lifetime investment. The liner is tucked under rocks around the edge and is all but invisible. His pond is spring fed. He did something I hadn't seen before. He put flat rocks on the sloping bottom of the pond and propped them up a little on the downside with smaller rocks, so the fish had places to hide from herons, kingfishers, and waterfowl.

A pond can have benefits not usually discussed. Ours provides water for our honeybees. It is amazing how the bees flock to the pond when there is no other source of water. They sip away all day and carry water back to the hive. Many other insects, not to mention birds, sip the water too, especially horseflies, which I could do without. They sort of dance over the water and every so often dip down to touch the surface. Every so often that is their last dance, too, because a bass can sometimes gauge the pest's descent to the surface perfectly and—bam!—no more horsefly. Dragonflies of exceedingly great beauty lay their eggs in the water or on the lilies and cattails. Their larvae eat mosquito larvae.

Does It Pay to Irrigate Pastures?

When I use the phrase "water in every paddock," I have in mind a third application: irrigation. Until 2003 I would have said that irrigation was too costly for typical pasture farming. In drier climates, irrigation is routine, even on alfalfa fields, so I have to conclude that someone has figured out how to make a profit that way. But in an area of frequent rain, like ours in a normal year, I would have said, probably a bit pompously, that the payback would not be sufficient to justify the cost. Now after two years of very dry weather (for us) and then a year of abundant rain in 2003, I am not so sure. In fact, I am quite unsure. What we learned in 2003 was that grass and clover will continue to grow right through late summer without a summer slump, keeping the livestock grazing almost as well in August and September as in May and June. Getting four inches of water in June, three in July, and two in August and September *that you can depend on* works a magic on pasture farming that convinced me absolutely that it will become a mainstream kind of farming. Furthermore, sufficient rain meant that we had plenty of stockpiled grass and clover for winter and plenty of moisture to start new seedings in August to get almost a year's jump over the following spring on new seedings. Since in many cases graziers are piping water to their paddocks anyway, would it not make sense to irrigate those paddocks, even when, as would be the case for us, we might need to use it only in July and August?

I was somewhat surprised when Sam Fisher, who owns Byron Seed Company, where I usually buy seed, said in a speech that he was irrigating his pastures in Indiana, where rainfall is the same as ours. I questioned him sharply, thinking he meant irrigation for seed production or trial plots, which would be a different financial equation than dairy farming. No, he irrigated his dairy farm pastures and thought doing so was practical. And the trick, he said, was to start applying water before the grass looked like it needed it. I suppose he wanted to be sure his pastures looked top of the line, since he is selling the seeds he was using on his own farm, but he was quite emphatic over the point that irrigation, even in his kind of climate, was something every dairyman ought to put the pencil to. And he was giving a speech in front of a group of farmers.

After I listened to him, I came to the following conclusion: if I were a commercial grazier, especially a dairyman, and I had the money saved rather than being forced to borrow it, I would install an irrigation system around at least some of my pastures. The increase in income would be more than what that money would make invested in CDs, bonds, or the stock market right now. Even if there were years like 2003, when I didn't need the extra water, it would still pay because irrigation equipment is a long-time investment, the cost of which can be spread over many years. If I didn't need to irrigate in some years, or needed to do so only once or twice, so much the better. Then the pastures would be giving top profit on their own, without wear and tear on my irrigation equipment.

For the smallest pasture farms, irrigation seems particularly enticing, although perhaps not commercially justified. The cost per acre I suppose would be higher, but the overall cost of supplying sprinkler irrigation to five to twenty acres would not be overwhelming. I am not going to venture estimating costs, because there are too many ways you could go to save money. To get an idea of cost, measure the borders of the paddocks where you would lay irrigation pipe, remembering that in most cases two paddocks could share the same pipe. Then go shopping to see what kind of plastic or aluminum pipe you can find used at a good price. Remember that you could most likely use a tractor you already own for a pump. You might have ponds big enough to use as a water source. The cost to ensure yourself that you have perfect pastures every year might be less than you think.

Imagine what a marvelous environmental paradise a pond-dotted farm would be, not to mention the awesome environmental benefit of being able to control the only weather risk in pasture farming: drought. What an awesome catalyst for an earthly paradise. And for the gourmet in all of us, really fresh fish from unpolluted waters, delicious beyond anything that you can buy. Don't forget frog legs, snapping turtle (delicious when fixed just like fried chicken), and duck à l'orange. Ah, heaven can't be too far away.

7

Pasturing Horses, Mules, and Donkeys

A common sight in rural America, and sometimes in not-so-rural America, is the barren lot of a quarter acre or smaller where a horse stands dejectedly nosing the bare dirt, looking for a blade of grass. The only reason the poor thing hasn't broken out of its mini desert is that it would rather go hungry than suffer the shock of an electric fence. There is a small shed at one end of the lot that offers comfort from the hot sun and flies, but not much, and where a bit of hay might be found when the horse's teenage owner is not too busy talking on the telephone. The barn is so full of manure that the horse can scratch its back on the ceiling. Most animal lovers seem to accept this woebegone scene with complete equanimity while they protest the idea of grazing two hens in the yard for fresh eggs.

Oh well. The situation does get worse—like two horses in the barren lot instead of one.

Horses are made for pasture farming. As Maury Telleen says in his classic book, *The Draft Horse Primer:* "Horses are basically pastoral animals; they are never happier or healthier than when on pasture." They do not even need the best pastures. Amish pasture farmers often graze their horses on paddocks after the cows have eaten the cream of the grass and have been moved to a fresh paddock. Horses especially fit small garden farms if for no other reason than that so many people keep a horse and don't appreciate the money-saving advantages of a bigger plot of ground to grow all the food the horse needs. Hay and grain can be expensive. A two- or three-acre horse pasture farm could pay for itself in time. I know of a case where a family already has enough land to do that but instead keeps a horse on a small barren lot and buys hay while there is an acre of lawn adjacent to the lot. I know of no more apt example of American lawn lunacy. In other cases, the riding horse is kept on the little barren lot while a large acreage next to it is planted to grain or is a mess of weeds in conservation reserve programs.

The general rule of thumb says that a horse should be fed about a pound of hay daily for every one hundred pounds of weight, in addition to a supplement of grain, which would not be necessary in most cases if the horse had access to really good hay. For a thousand-pound horse, that would be ten pounds per day minimum or something over a ton and a half a year. A ton of *good* hay will cost you two hundred dollars or more. Trying to save money with cheaper hay will cost you more in the long run. So hay for one horse per year would cost around three hundred dollars. Most horse owners will buy at least another hundred dollars of an expensive grain supplement whether they need to or not, so that's roughly four hundred dollars a year. And buying in hay, you risk new weed seeds, harmful molds inside the hay bales, or hay with blister beetles in it. Four hundred dollars is not a bad income from two to three acres of pasture. If the plan described in chapter 4 is followed, the two acres could be divided into four or even eight paddocks. A paddock or two could be planted to oats every year along with a legume. The oats could either be grazed when ripe or cut and stored like hay to provide a horse with an excellent supplement and to save buying grain. Both the oat heads and the oat straw are excellent feed for horses. Fed this way, it used to be called "sheaf oats." On the other paddocks, a cutting of hay or two during the June season of surplus pasture would provide, along with the sheaf oats, winter and supplemental summer hay. A first cutting of hay should yield about two tons per acre. Oats planted in August on a garden plot after an early vegetable has been harvested will provide pasture in winter.

It can be argued that when workhorses are being worked hard, they may require some concentrated energy food such as corn, but most riding horses and cart horses, as well as workhorses when not working hard, do not get enough exercise to justify purchased grain rations. A little oats provided as described above, a pound or two a day at the most, is enough. In fact, the tendency is to overfeed riding horses. Then, one fine day the novice owner begins to wonder if leather shrinks, because the saddle doesn't fit on her horse anymore. Just look at those magnificent wild horses of the West. They flourish on rangeland that looks as if a horse would have to run half a mile between blades of grass to keep from starving to death.

Horses will graze all winter even in the North if there is any plant life at all available above the snow. I have seen them in deep snow browsing Japanese honeysuckle vines, a fencerow and woodlot pest weed in the mid-South and lower North. (This weed is fairly nutritious, Bob Evans tells me. He believes honeysuckle can provide practical emergency pasture for cattle, too, during periods of heavy snows. See the chapter on pasture weeds.) Amish farmers overwinter their horses in fields of cornstalks after corn harvest. Horses have been overwintered on oats straw. Here in northern Ohio, my sister has always kept her riding horses out on pasture almost every day of the year, saving a small patch of heavily sodded fescue for those days in late winter or early spring when pastures might otherwise be too soft for horse traffic.

A high-tensile fence with top and middle wires electrified or a woven wire fence with an electrified strand above it is best for horses. Without electric fencing, a hungry horse will ride down even barbed wire if it fancies there is some delicacy on the other side. The traditional cattle fence of three to five strands of barbed wire is okay for cows, I suppose (I don't like it), but horses are especially prone to stepping over the bottom wire and getting tangled in it. That can mean a badly lacerated leg. I don't see any sense in making fence out of strands of barbed wire now that high-tensile electric fence is available.

This is not a book about horse health, but some problems are connected directly to pasture feeding. Unusually warm, wet weather in spring can generate very lush grass, upon which a horse may founder just as it will on too much grain. This is especially true when horses have been penned indoors on very lean diets all winter and then are suddenly given the run of very lush cool season grasses like bluegrass or ryegrass when the dew is still on the pasture. That's like locking an alcoholic in a bar. Farmers allow draft horses only gradually on such lush grass. For riding horses, it is better in early May to keep a horse on a small pasture that has been kept nibbled down all spring. Horses work well with sheep in this regard, because where sheep have been grazing in early spring, the grass is not likely to be lush enough in May to founder the horse. The chance of grass foundering diminishes as the grass hardens a bit and starts going to seed in late May.

A basic reason why horses are good candidates for pasture farming methods is that they generally are not kept to produce meat or milk for human consumption (they are, in special circumstances in other countries), so there is no need to overfeed them extra protein and energy like farmers traditionally do in milk, pork, or beef production. Where bluegrass and white clover grow well, which is most of the country, they can form the basis of the horse's diet most of the year. Indeed, many horse owners prefer bluegrass, especially on the calcium-rich soils of the mid-South. That's why racehorse country has tended to be centered in the Kentucky bluegrass country. In spring and again in the regrowth in the fall, bluegrass can actually contain nearly as much protein and calcium as legumes, although some authorities say some of that protein is not available. I am leery of all conversations about what nutrients are or are not available. I've listened to too many nutritional preachers trying to sell their particular brand of supplement to believe any of them. All you need to do is watch your horses. If they are sleek and frisky, they are almost always fine. For supplemental pasture during the cool season grass slump in a dry summer, a seeding of oats (three bushels), red clover (eight pounds), and timothy (five pounds) will make good pasture. Planted in August when rainfall is adequate, it will make good horse pasture in winter, too.

Mules and donkeys thrive on the same kinds of pasture as horses. I have read that they will eat nodding, or musk thistle (Carduus nutans), and the thorny horse nettle (Solanum carolinense), but most graziers disagree.

Donkeys are sometimes used as guard dogs for sheep and goats with varying results. Be sure to introduce a new donkey to your pasture community slowly—with a fence between them for a few days. That's a good rule for introducing any new animal. Two donkeys are not better than one: they will form their own little society and ignore the sheep or goats. One alone will bond with the flock. A horse can be just as protective of a flock of sheep, an observation that gave us considerable amusement on our farm. When we first started out, with two sheep and a riding horse called Misty, the sheep would run to the horse and stand right under her whenever a dog entered the pasture field. If the dog came near, Misty took out after it with bared teeth. Friends of ours say their mules will do the same thing.

Misty also fancied herself something of a sheep dog. When she wanted to go to the barn, she would drive the unwilling sheep ahead of her. I guess she thought they were her colts. My mother had a riding horse that would try to steal baby calves from their mothers. The horse would get the calf in a fence corner and not let the mother approach or the calf get away. This can be a problem with donkeys, too, I understand. They will try to get a very young lamb away from its mother and keep it from nursing. Another interesting fact about donkeys: they usually are gentle and forgiving around children and so make good pets for youngsters.

Not far from our farm lives a most unusual farmer, Jack Siemon, who has made a good living raising mules, or what are known in the trade as mammoth jacks. I have written about him in detail in a previous book (*Living at Nature's Pace*), but it is proper to mention him here because he was one of the first modern grass farmers in our part of the country, practicing the principles long before that term became popular. Jack Siemon's genius early on was in recognizing the advantages of pasture farming over till farming and then teaming it with a grazing animal that he could sell off the pasture at retail prices. Surprisingly (to most of us, not to Jack Siemon) a lively demand for his jacks developed worldwide. Not all the world's farmers have gone tractor crazy, not because they wouldn't like to but because they can't.

Age has now forced Mr. Siemon to cut back on his farming, but ten years ago, when he was in his seventies, his farm of only 180 acres was a remarkable sight. Almost all of it was in lush legume pastures. Amidst the plowed thousands of acres around the farm, you could spot it easily in November or December from a jet plane high in the sky, an emerald shining in an earthy ocean of brown. The farm was a model of pasture farming methods. Mr. Siemon grazed the legumes and grasses in rotation, moving the animals from one field to another and then back to the first to start over. One field he let grow up in late summer for winter grazing. After years of grazing he didn't have to fertilize these fields or spray them with weed killers. He did not have to reseed either, unless he wanted to try some different forage crop. What the animals could not eat in June, he made into hay.

And what a menagerie of animals he kept: eighty head of mammoth jacks; a dozen draft horses; a couple of light harness horses; a few dairy cows and calves; a little herd of beef cattle; a flock of sheep; turkeys, ducks, and many kinds of chickens, geese, and guineas. Jack Siemon was a master husbandman. By careful breeding, he was improving all these animals and selling the offspring. He reckoned that there were about three hundred farm animals on his 180 acres. He kept two employees busy most of the year. Here was a farm of only 180 acres providing a living for a farm family and jobs for two hired helpers. If all the farms in America were like this one, there would certainly not be any unemployment problem. And farmers would certainly not have to be subsidized to keep them in business.

Oddly enough, the donkey is the workhorse of choice in some parts of the world. At Horse Progress Days, an annual event that was held in 2003 in Mt. Hope, Ohio, I was talking to Eric Oganda from Kenya, Africa. I assumed he was there because he was interested in workhorses. To my surprise (there are a whole lot of things at Horse Progress Days that will surprise you), in Kenya the donkey is used more than the horse for draft power. Horses are too big and too expensive for Kenya's meager resources. Mr. Oganda was looking for draft horse harness and machinery that could be scaled down for donkey power.

At Horse Progress Days, people (about ten thousand of them) from all over the world come to see the amazing innovations being perfected in draft animal power. On display were inventions that allow draft animal agriculture to compete quite handily with tractor power. Horse farmers now commonly use what are called hitchcarts behind their horses. For example, hitchcarts are being manufactured that are equipped with ingenious hand-operated hydraulic systems, power takeoff assemblies, and three point hitches. Any equipment made for smaller tractors can be attached to the hitchcart and pulled by horses. Horses pulling hitchcarts still cost less than tractors. A recent report from an Ohio State Extension agent who works in Amish country concludes that Amish farms are quite a bit more profitable per acre than tractor farms. It gives me pause for thought. Long pause. As horse farming moves forward in the twenty-first century and pasture farming gains in popularity, could there be a rebirth of interest in horse, mule, and donkey power, the animals most suited to the grazing life?

8

Sheep on Pasture

Sheep are the easiest animal to raise commercially in a pasture farming system. That means they are an even easier project for noncommercial pasture farmers. They are easier than beef cattle to finish on grass alone. They are very efficient foragers. A healthy human can lift them for examination or loading onto a truck. Can't do that with cows. And sheep can run almost year-round on the fields without risk of overly pugging the soil surface during thaw times.

For those looking for a low-cost way to work into commercial farming, sheep offer interesting market opportunities. For example:

1. Lamb prices are generally on the profitable side, if only a little. But the wide gap between market price and grocery store price, to say nothing of restaurant price, offers an opportunity to process the meat at local butcher shops and sell it directly to customers.

2. Spurring this market is the increasing number of people with roots in Islamic culture (and other ethnic groups, although I hate that word—we are all "ethnic") who prefer lamb over beef or pork. Some religious groups like to come directly to the farm and perform their ritualistic slaughtering and butchering tradition. Tuned-in sheep producers try to time marketing schedules to Islamic holy days, when the demand

for lamb goes up. Every sheep producer should hang around the markets where they plan to sell. I sit in the upstairs restaurant at the Bucyrus, Ohio, stockyards and ask questions while I eat all that delicious fattening food. It is a better school, and a whole lot more enjoyable, than agricultural classes at a university. The language is a lot more colorful, too.

3. There is also a growing interest in hair sheep—not only because they don't have to be sheared but also because many people, especially those of Hispanic origin, prefer the taste of the meat over that of wool sheep.

4. The wool market also seems to be stirring a bit again; this year (2003) our Corriedale wool, at thirty-five cents a pound, paid for the shearing for the first time in four years. (Only the very fine Merino-type wools are profitable at the moment.) But there are ample niche markets for shepherds willing to process their wool and sell it themselves. Scattered all over the countryside are small flock owners who spin and weave fabrics and blankets and woven artworks from their wool for good part-time cash. One couple I know sells comforters padded with their wool.

5. Some shepherds have gone into sheep dairying to produce Roquefort-style and other cheeses for specialty markets.

All of these farming specialties can be pursued without turning one furrow of dirt. Sheep don't need corn. When our pastures are ample, mine won't even eat it.

There are many breeds of sheep, and they are all the best—just ask the breeders who sell them. There are fine-wooled sheep, like Merinos; coarse-wooled sheep raised for meat, like Suffolks; and hair sheep, like Katahdins. Some meat breeds also have fairly fine wool, like Corriedale. There are colored-wool sheep (browns, grays, blacks), long-wooled sheep, sheep with long horns, and sheep without any horns. Some breeds are touted for multiple births, like Finnsheep, though, as I've said, I think triplets and quadruplets are a great pain to be avoided whenever possible. The affliction of agriculture is the desire to produce more. Once the market is saturated, producing more is lunacy, yet we go on doing it. If most farmers were content with twins or even singles (or with corn at one hundred bushels per acre or with grazing one cow and calf per acre, etc., etc.), then a few of the more ambitious ones could profit from producing more. But of course, any method that increases production must be adopted by everyone who wants to survive in the marketplace. The market becomes oversupplied, dog-eat-dog economics prevails, and nobody profits.

Breeders bragging about the relative merits of one breed over another remind me of teenagers bragging about their cars. Variation within a breed is more important than variation between breeds. In other words, whatever breed you choose to work with is not as important as the number of animals of that breed into which you can develop the traits all shepherds want. In every breed there are ewes that are better mothers and better grazers, and those that are more apt to get pregnant, more apt to bear twins, more apt to have trouble-free birthing, and more apt to resist internal parasites. These are the traits you cull and select for. After years of upgrading, a flock becomes valuable, no matter what the breed.

If I were to start today, I would probably choose a smaller breed than Corriedale, not because smaller breeds are thriftier grazers, which I doubt, but simply because now that I am older, I can't handle bigger sheep as easily. We got into Corriedales because a nationally respected Corriedale breeder and show judge, Al Kin, lives conveniently just down the road. He's a great neighbor, and when we were getting started he was willing to bring me a ram when I needed one and take him back again after breeding season.

A most important part of husbanding sheep, whatever breed you choose, is to try to maintain a closed flock after you get started. Buying in sheep can mean buying in problems like foot rot. You have to bring in rams, of course, and ewes if you are trying something different, but otherwise, breed up from within your own flock.

In managing sheep, I like to keep in mind a traditional folktale that I first read in Bob Kaldenbach's *Rural New England* magazine, now defunct. I've told it before, but it can't be repeated too often:

Once upon a time, an ambitious young man decided that he wanted to make money raising sheep even though he didn't know much about them. Being a practical man, he asked the sheep. After all, who would know better than a sheep how sheep should be raised. "Simple, really," the old leader ewe of the flock told him. "All we need is a bite of grass, a sip of water, and a pinch of salt." As long as the new shepherd followed the sheep's advice, things went well for him. He even made a little money. But if the sheep were content, the man was not. He asked them again: "Are you sure you don't want something more, a barn perhaps to sleep in, better kinds of grass, some grain, a little protection against wolves?" To which the sheep assured him again that all they needed was a bite of grass, a sip of water, and a pinch of salt. But their master would not listen. He drove himself into a frenzy of debt and stress to provide his sheep with all the things that he was sure they needed, meanwhile complaining that sheep were costing him money. The sheep responded only by reminding him once more that all they needed was a bite of grass, a sip of water, and a pinch of salt. At long last, broke and tired, the man understood: he himself had been the cause of his problems, not the sheep. "Sheep don't need much, it is true," he admitted. "Most of what I've done is for me."

Raising sheep on pasture is not that simple, or rather is that simple if you are not striving to make a big profit. Otherwise, sheep, at least fattening lambs, need more or less constant access to top-quality pasture to gain weight properly. But sheep on the best pasture also need to have some poorer roughage, a fact not often mentioned. I learned that from Dutch farmers who feed old hay or straw routinely when their animals are on lush pasture. I used to try very hard to get hay in without rain. Now I use lower-quality, rained-on hay, in stacks, to balance out lush grass. The sheep will go over and eat on the stacks occasionally on their own. They need that fibrous roughage to keep from getting too loose. At least I think that's why they do it. Maybe they are just being contrary. The point is that the grazier can go overboard (and some do) trying to keep lambs (or milk cows) exclusively on the cream of the grass crop.

Sheep on lush spring pastures may succumb to grass tetany, which is caused by a deficiency of magnesium and calcium. It is a rare occurrence. It has never occurred on our farm. I think the reason for our good fortune is that our soils are fairly well balanced and fertile and contain enough magnesium and calcium. Also, our sheep are out on pasture in early spring as the grass begins to grow. They become accustomed to lush growth gradually as it comes on. There is never the shock of suddenly going from very hungry to gobbling a rich pasture fast.

Foot rot can be a problem in sheep. It is said to be highly infectious and if you know one of your sheep has it, you probably should separate it from the flock. (Notice how I salted that sentence with "said to be" and "if" and "probably"—I don't think a whole lot is known for sure about foot rot, and I know for sure I don't know much.) When a sheep is limping and the sore hoof looks a little puffy and has a rank smell, it *probably* has foot rot. Some shepherds think sheep on marshy land are *more apt* to get it than on well-drained pastures. I doubt that the bacteria that cause foot rot care whether the sheep lives on dry or wet ground. But these bacteria thrive in the absence of air, and a wet hoof *probably* blocks air and sets up a habitat more to the bacteria's liking than a dry hoof provides. About the only thing you can do is pare back an infected hoof very closely so the wound gets well aerated and administer an antibiotic if your vet prescribes one. A footbath of zinc sulfate is another standard ministration.

I feel uneasy trying to give advice about foot rot, because I don't think we've ever had a case in thirty years. If we did, we didn't follow the proper directions to cure it, because we didn't know what was wrong. Quite commonly one of our ewes comes up limping. I panic. But after many years with ailing sheep, my first rule is DO NOTHING. It is amazing how often that works. The ewe will not be limping the next day. When I examine a ewe that persists in limping, it is usually a curled up hoof that I should have trimmed before it got that way. I've never figured out hoof trimming, either. Some ewes need it; some never do. A few times when I thought I smelled hoof rot, I trimmed back the hoof very close and cleaned out all the dirt. But I didn't use any medicine, and I didn't call the vet. In all these cases, the ewe quit limping in a couple of days. Perhaps I'm just lucky. Perhaps moving sheep regularly to new paddocks does not allow foot rot, if that's what it was, to get established and spread. Perhaps, perhaps. The world is one grand perhaps.

Bloat is another notable health problem associated with grazing sheep. There are all sorts of theories about what actually causes bloat, a favorite being that clover growing very fast in a warm, wet spring develops a calcium deficiency or imbalance that makes conditions in the sheep's gut congenial for bloat. Perhaps again. And then again, perhaps not. What no one argues is that if sheep (or cows) eat very lush legumes very fast, they may bloat. Gases get trapped in the rumen, the gut swells, the lungs get compressed, and the animal dies, sometimes very quickly. But most of the time, a sheep will only bloat mildly and will be fine. When the case is serious, the drastic way to save the animal is to puncture the swollen rumen and let the gases out. If you don't make the puncture in precisely the right place—that is, on the animal's left side in that hollow in front of the hip bone—she might die from the wound, not the bloat. While you are killing her, another bloated sheep that you could have saved just by getting her up on her feet might die too. Another old shepherd's tale is to shove a length of hose very slowly down the sheep's throat about three feet to clear the esophagus and let the gas out through the hose. Yeah. If you are lucky enough to make that work, stand back because that gaseous effluvia might come out with force. Puncturing the rumen through the flank also causes the gases and sudsy fluids to come spraying out on you. Worse than getting sprayed by a skunk. I once punctured a bloated cow with a pocketknife, and that's how I know. She lived, but I almost didn't.

I don't think a novice should try any of these emergency efforts if the animal involved is a sheep, unless the sheep happens to be a very valuable one. (On the other hand, if you don't try it, you will always be a novice. Life is so complicated.) Better to get a veterinarian if there is time, which there often isn't. (I just said it—life is so complicated.) Prevention is, in my opinion, the only way to handle bloat, and when you fail to prevent it, accept the possible consequences. A dead sheep. Actually, they don't always die. If you can lift them up and keep them on their feet, they sometimes survive. I've seen cases where straps were passed under the animal to form a kind of sling to lift it and hold it up until it worked the gases out and gained the strength to stand on its own. A front-end loader makes a handy lifter.

Bloat is not difficult to prevent. I have never had a case in sheep, and the one in a cow happened when she got into some high-moisture corn. Alfalfa is most often the legume that causes bloat, because it is the legume that most often grows thick and lush. Always let the animals fill up on good hay before turning them into legumes so that they are not inclined to eat too fast. Make sure there is a grass like timothy or orchard grass with legumes. If a sheep or cow can eat grass and legume together, they are not apt to bloat. Never turn sheep into lush June alfalfa or red clover. Never turn them in on dewy wet legumes, even if the stand is not lush. And, above all, never turn livestock into a legume that has just been hit hard by frost. Wait three days. I do turn livestock in on second- and third-cutting alfalfa or red clover after the dew is off. It is not as lush as first cutting. But always the animals get hay first. After the animals become accustomed to eating a legume, they can be left in the field. Immediately after turning the flock into a fresh paddock, I hang around for awhile, walking among the sheep, keeping them moving so that they don't just stand there and gulp down great gobs of legume. They have never done that, because they aren't overly hungry after eating hay beforehand. They will graze in their own common-sense way, which is to nibble some clover, then some grass, then go over to the margin of the field, which is all grass, and eat some of that.

If you want to feel safer about bloat, there are salt and mineral blocks and other supplemental feeds that are supposed to protect against it. I wouldn't want to rely completely on these safeguards, and if you follow the instruction above, you won't need to. I've never used a bloat preventative.

Internal parasites (worms) are the limiting factor in pasturing sheep by intensive grazing methods. Permanent pastures harbor worms, and the eggs can remain viable for over a year. Rotating pastures therefore does not get rid of worms unless more than a year elapses between grazings. But rotations with intervals of more than a month do help some. Mowing or making hay from a paddock helps some, too. Cultivation, even a light disking, can break a worm cycle. But unless you have a huge ranch, or have enough land to graze half of it one year and the other half the next, back and forth annually, or graze half with cattle and half with sheep and switch every year, you will need to worm sheep. Ewes need to be wormed especially around lambing time, and the lambs shortly after. Some shepherds keep ewes and new lambs in a dry lot for a while, feeding them hay and grain, simply to avoid the burgeoning worm population in the warm, moist spring pastures. But pasture lambing is not all that grim. If you can use a paddock for lambing that has been recently cultivated or recently hayed, it may be relatively free of worms temporarily. A second worming of the flock should take place about a month after lambing or, if worms have been a problem, three weeks later and again three weeks after that. But many years we have gotten by with one worming, and no year yet have we wormed more than twice. We use two different wormers when we treat the sheep twice. I credit our ability to control worms so far to mowing the pastures regularly and turning the flock into the "wild" paddock, where there are walnut seedlings and other plants to browse that, at least according to folk medicine, act as vermifuges. Also, we have had a lot of dry hot summers and open winters lately, both of which are hard on worms. In any event, keep the flock penned off the pasture for a day after worming so that the live eggs and larvae in their manure don't re-infect the pasture.

Another effective worm control is to graze different kinds of animals together. Cows and sheep are not generally affected by the other's internal parasites, so the cows eat and destroy sheep worm larvae and eggs and sheep do the same favor for the cows. Chickens, ducks, and geese roaming the pasture like to scratch in the droppings of cows and sheep and eat the worm eggs in them. An ancient worm preventative in merry olde England institutionalized this phenomenon. The pasture was divided into thirds, a portion each for weaned lambs, ewes, and cattle. The third, which the lambs grazed one year, was given over to cattle or to haying for the whole next year. It was called the British Rutter Ring System, and evidently it worked quite well.

What really brings on the internal parasite problem is overcrowding the pasture. I have yet to talk to a shepherd who did not agree with that. But most of them go out and overcrowd anyway. Money moves mankind more than moderation.

Sheep do need that pinch of salt, like the folktale says. I vary a block of plain salt with a block of mineralized salt. Mineral blocks usually contain copper, which is not good for sheep, so I'm told. I really don't think I need to feed mineral blocks at all when the pastures are good and the animals have access to a wide variety of forages, including tree leaves and weeds. It's sort of like taking vitamins. I don't take them because I think my diet contains plenty of vitamins. But I keep on feeding mineral blocks to the sheep just in case I'm wrong. I guess I care more about the sheep than I do about myself.

And of course sheep need water. Water is important beyond the fact that all living things need it to survive. Water helps keep sheep or any animal or human in good health. If the water is clean, the animals will drink more of it, and the more they drink, the better for them. I don't know of an easier way to get supplemental fertilizer on a pasture than livestock drinking from the pond and then urinating on the pastures.

Wild dogs or domestic ones, coyotes, wolves, and eagles are threats to sheep. Tame dogs are the worst because there are so many of them. They aren't necessarily vicious. The sheep run, even from friendly dogs, and so the dogs chase, and if it's a hot day, the chase may go on too long, and the sheep collapse from heat exhaustion. Coyotes often get the blame for what dogs do. In the "old days," like no more than forty years ago, a farmer had the uncontested right to shoot dogs chasing his sheep. If you do that now, you better be sure there's wool in the dead dog's mouth when the sheriff checks out the shooting. Even then, you might get prosecuted. Modern society lives so far removed from the realities of food production that a pet dog has more status than a whole flock of sheep. And heaven help the rancher who shoots a bald eagle carrying off a lamb. In today's fever of patriotism, you might go to jail even though bald eagles are on the increase in population, at least here in Ohio. Actually, to me it would be worth a lamb to see an eagle carry one away.

I am often asked if foxes kill lambs. In my experience they do not. Red foxes are part of our pasture wildlife community, and I have seen them mingle right with the sheep with no apparent interest shown on the part of either. Red foxes are beautiful, and I gladly donate a few chickens a year to their health and welfare. That's the problem. This year a few chickens; next year a whole coop full.

Electric net fencing will keep coyotes and dogs at bay most of the time. Since generally coyotes go after very young lambs, you can coyote-proof the pasture that you use for lambing and reduce the chances of predation considerably. Small pasture farms, close to house and barn, have fewer problems because of the proximity of the shepherd. Large pasture farms with paddocks out of sight of barn and house, and out of smell of humans, are the ones more at risk.

Since we are around the flock much of the time, we have lost only two lambs to coyotes or dogs in twenty-five years. I think coyotes were to blame because the stomachs were opened and the intestines pulled out, a trademark of coyote predation. Dogs rarely eat sheep. They just enjoy running them to death. If you see your flock bunched together, with the ewes on the outside facing outward with lowered heads, dogs may be bothering them. Move fast. Coyotes try to slip slyly into the flock without alarming the animals and then grab one. There is almost always some partial carcass left behind. One time, a lamb completely disappeared, so I assumed a bald eagle carried it off. Coyotes or dogs might be able to get through most of our fences, but not carrying a lamb. If I hear coyotes howling at night, I walk the fields shining a flashlight about (coyotes don't like flashing lights, I've read) and perhaps shoot off the rifle to scare them. I don't mind saying that I use every opportunity to urinate in the paddocks, too, hoping the odor will scare off the coyotes.

As far as caring for sheep in general, one learns over the years that sheep love to die. In addition to grass tetany and bloat, there are a host of other diseases. Even if you or your veterinarian can keep your flock safe from them, you are still not home free. A few sheep will find new and imaginative ways to end it all. One of my sisters kept sheep in her orchard. Her children tied a rope swing onto one of the apple tree branches. A ewe managed to hang herself on the swing. I had a ewe commit suicide by eating poison hemlock, even though the plant is so absolutely bitter that livestock will seldom eat more than a taste. Thought she was Socrates, I guess. My saddest story concerns one of my best ewes (always the case, it seems) who drowned herself. I had cut a hole in the ice on the pond for the sheep to get a drink. The hole was not big enough for a whole ewe to fall into. But with the genius of the suicidal, that ewe managed to slip as she drank, lunge forward, and ram her head into the hole. She was unable to gain footing so she could back out.

I do not think it pays to use extraordinary and costly methods to save sheep. The most successful pasture farmers take a rather ruthless view. The ideal ewe is one that goes out on pasture in the spring and comes back when the snow flies, with a nice market lamb or two at her side. Everything else is culled by nature if not by the shepherd. The shepherd likes buzzards very much. They clean up the sheep that nature culls. The ewes that come back to the barn halt and lame or without lamb go down the road no matter what the pedigree. That sounds hardhearted, but most commercial flocks and cattle herds in the West are managed that way. More time and money is spent caring for weak or unhealthy sheep than the sheep are worth. And invariably they will die anyway, or they will birth lambs that will die anyway. You can let sentimentality cloud your judgment, but all it will mean is more cases where you have to let sentimentality cloud your judgment. And you really do want a sheep to die occasionally to keep the buzzards around.

When you start out as a shepherd, likely as not, about 15 percent of your flock is going to give you trouble. You can work yourself to death trying to save that 15 percent, and the result will be that you will continue to have 15 percent giving you trouble. Half of that 15 percent will die or amount to nothing anyway. The other half pass their problems on to their offspring. But, yes, do the ordinary, practical things like pulling lambs that need a bit of a gentle twist to ease them out of the birth canal. Do watch out for maggots burrowing under the wool. Scrape them out and treat the wound with healing powder and insecticide. That job will separate the true shepherd from the make-believe. (By the way, don't douse the wound with gasoline. Every time I have done that, the lamb died. Smear petroleum jelly on the affected area, taking care to get out all the maggots first. Then spray insecticide. As the wound heals, I have found I could use one of those Raid cans that spray a stream of insecticide, useful in knocking out wasp nests. If I see those blasted blue flies that cause maggots buzzing around the wound, I can give the lamb a squirt without having to catch it.)

If there are triplets in the womb, twisted and entwined around each other, your chance of saving those lambs is not worth the cost of the vet's visit, in my opinion. Don't think yourself a failure when you lose them. Consider yourself lucky to save the ewe. And then sell her as soon as you can. Many shepherds go to great lengths to select ewes that throw triplets, believing that three lambs per ewe is better than two. My experience is that triplets don't pay. More than half the time, one of the triplets will die or not thrive. The goal of pasture farming is a ewe who can take care of herself and her offspring. For me, that usually means one or two lambs. Triplets will force you back to barn and pen farming, not pasture farming.

Invariably, a few ewes who birth twins or triplets will abandon one of them. This happens less when pasture lambing, because the ewes can isolate themselves from the flock to have their lambs. The firstborn does not wander off to mingle with other lambs or ewes while the second is being born, as happens in the barn if you don't get mother penned up before birth. Other times, a ewe just gets it into her head that she will allow only one lamb to suckle. All those tricks you read about to persuade the mother to take back her rejected lamb hardly ever work. They are not worth the effort. If you won't take my word for it, you will just have to learn by the sweat of your brow.

Instead, you raise the lamb on a bottle. The only real benefit of doing that is to make yourself feel better as a shepherd and to give your children something meaningful to do. Otherwise, bottle lambs often cost more than they are worth. I have a shepherd friend in North Carolina, Brian Knopp, who believes that bottle lambs are the curse of the flock. If you save them for breeding, they may pass on to their offspring whatever it was that made them a bottle lamb in the first place. If you raise them for the lamb market, they usually gain weight so slowly on pasture that they bring a low price at market time. A bottle lamb does not thrive on pasture because that fake milk you are feeding it just isn't the same as the security and nutrition of a lamb nursing mother. (I would bet anything this is true of humans, too.) If a lamb is going to get maggots, it will probably be one that you are bottle feeding. If a coyote or a dog gets a lamb, it will be the bottle lamb because it has no mother to protect it. If you try to feed the lamb only twice a day, it will get so potbellied you won't want anyone to see the ugly thing. Even if you feed it three or four times, as you should, it may still get potbellied, and it will still gain slowly. You are spending time that would be more profitably spent elsewhere. You can put bottle lambs on self-sucking buckets so they can drink whenever they feel like it, a little at a time, as they do from their mother, but that fake milk is expensive beyond the gain in weight it produces. And if the weather is warm, the flies will drive you and the lambs to distraction.

At least with self-feeding, the lambs don't become pets. Pet lambs are lots of fun for awhile, but they always want to be where you are. Only the very best fence can stop them. Sometimes when you try to wean them from the bottle, they will stand at the gate and bleat until they nearly starve to death. They have transferred their flocking instinct to you. They think they are human. They don't know how to be sheep anymore. If they survive and become ewes, they will always be in your way trying to attract your attention as you work around the flock. If an orphan lamb dies before you can get it trained to the bottle, do not feel sad. Just mark the mother for shipment. Nature has saved you from losing money and just may have helped you improve your flock.

If you are an old softie, like me, you will go on feeding bottle lambs and losing money at it. Being an old softie is not the worst of faults. We have even kept bottle lambs in our house, or did before we started pasture lambing in the spring after the zero weather had passed. My wife put diapers on a couple of them, for heaven's sake. Yes, cut a hole in the diaper for the tail. Diapers cost about as much as the lambs were worth. But the lambs were very entertaining as they raced around the house or snuggled up in our laps to watch television. In the latter case, I almost always fell asleep first, which proves that television is aimed at an irrational audience. Just remember that cuddly little lambs grow into hundred-pound miniature horses, and they will still want to come into the house.

Traditionally, lambs are weaned at about 70 to 80 days, but grass farmers often prefer 90 to 120 days, by which time the lambs self-wean. Weaned lambs, separated from their mothers, can be fed diets higher in protein than the ewes need and will thereby gain faster. Most graziers run the weaned lambs alone on a paddock of new grass and then let the dry ewes clean up the less desirable plants while the lambs go on to a new paddock. This separation also allows rebreeding of the ewes without bothering the lambs being fattened for market.

With our small flock, we see no reason to wean our lambs. They wean naturally in about four months and continue to run with the ewes until we ship them to market. I think the extra milk and emotional security they get from staying with their mothers makes up for the supplemental high protein diet that weaned lambs are fed for fast gains. Perhaps our lambs do not gain weight quite as fast or as much, but I like to wait nine months—after the market is starting to rise again in November—to sell them anyway. In fact, with low-cost pasture farming, I have waited until January or later to sell lambs and more than made up in price the extra pasture and hay I fed.

In grazing, sheep favor certain parts of the pasture over others. First they eat all the best-tasting plants, like early bluegrass, white clover, and young timothy. For reasons I can't explain, they will generally graze down poorer hillsides harder and faster than more fertile bottomland soils. Jim Gerrish, the pasture scientist I trust the most, says it's because nutrient density is greater in slower-growing plants. Within a paddock that has both kinds of soil situations, it is impossible to graze the whole area evenly, as the grazing manuals instruct. Sheep will gnaw down bluegrass and clover on the hillsides first, then the same plants on the better soil, then turn to ryegrass, fescue, orchard grass, and other coarser grasses. They will go back daily to the poor hillsides to nibble even dried-up bluegrass before filling up on the greener, coarse grasses available.

When moving the flock from one pasture to the other, the grazier takes into account all these details. Overall, the pasture should not be grazed down less than about two inches. The sheep should remain until they eat at least some of the less tasty plants. A happy medium is difficult, if not impossible, to attain. They will overgraze where the grass is more to their liking and undergraze where it is less to their liking. I think it is better to err on the side of overgrazing a little than undergrazing, just to aid in weed control. My experience is that a hillside gnawed too close, according to the manuals, will recover just fine when it rains again. But weeds not controlled mean more weeds shading out the clover. I usually clip the paddock after I have moved the animals to a fresh paddock. Doing that may not be necessary all the time, but I do it anyway. It keeps the regrowth uniform in height, and I think that is important.

There are many variations of rotations and forages. Richard Gilbert in southern Ohio, mentioned in chapter 3, has tried some unusual ones. He divides his pastures into three main categories, spring/early summer, high summer, and summer/fall. For all of them, he wants a strong white clover component. In addition, he currently likes, for spring/summer, a vigorously tillering orchard grass like Tekapo along with Kura clover (which is technically not a white clover but similar to it). For high summer, when cool season grasses fade a bit, he suggests Marion lespedeza and Red River crabgrass. For late summer/fall he likes endophyte-free tall fescue, alfalfa, and birdsfoot trefoil. Depending on circumstances, any of these pastures might also be grazed at other than their main rotation time. He lambs on the spring pasture, then rotates to the summer/fall pasture. The lambs are weaned then and put on the summer lespedeza and crabgrass, where they finish in a relatively parasite-free environment (because tillage is involved in getting the crabgrass germinated well and tillage breaks the worm cycle). Then for the ewes, it's back to the other two pastures for shorter rotations while saving some of the summer/fall for winter pasture.

In a situation where land is limited, Mr. Gilbert says the crucial management decision involves letting enough paddocks grow in August, September, and October to stockpile for winter-long feeding. If he doesn't have other paddocks to pasture at this time, he feeds hay in the fall rather than winter, a highly unusual practice. "It makes sense for me," he says. "It is good for the grass to be growing during the fall when it is building up vigor in the roots for next year's growth. Hay purchased in the fall is usually cheaper than hay purchased in winter. Also, when I put my sheep on the pasture I saved for winter stockpile, it flushes them for breeding beautifully."

Spend as much time as you can watching your animals graze. Study how plants react to grazing. For example, horse-nettle is a problem for us because the sheep won't eat it. But to my delight, I noticed that they do eat the blossoms. So the weeds don't go to seed and can spread only by root rhizomes. In time, grazing the blossoms along with mowing to develop a thick vigorous grass sod lessens the problem.

Don't spoil your sheep, especially the ewes after they are no longer nursing until they give birth again. Sheep know how to follow you around and bleat pitifully, as if they were starving to death. They, of course, want the most lush, tender grass and clover that you can give them, and, like humans, they always think it is somewhere other than where they are currently grazing. Make them clean up their plates. If you are feeding supplemental hay in winter, make sure they are still hungry enough after eating the hay to go along the fencerows and other areas of rough grass, where, if you look closely, there is also some short greener growth underneath. Sheep will clean that all up once they realize they aren't going to get all the good hay their little hearts desire.

Traditionally, most shepherds, like Mr. Gilbert, "flush" their ewes right before breeding them. The ewes are turned into a lush pasture saved for that purpose, or they are fed grain. The theory is that a jolt of food very high in nutritional value will mean more multiple births. The efficacy of flushing is well documented, but, being a contrary farmer, I wonder exceedingly if the right cause and effect are in play here. I think flushing seems to work because the sheep are generally on poorer pasture up to the time of breeding, so of course turning them into lush pasture can have a beneficial effect. But I think ewes should be on good "flush" pasture at all times if possible. My bias is that a ewe is going to have twins based on genetic inheritance, not nutrition. Understand, I am in the decided minority in this matter. I just can't be convinced that diet has anything to do with multiple births. I actually do turn my ewes into fairly nice red clover in November when I breed them, but if I thought this was going to encourage triplets, I'd discontinue the practice. In my experience, my ewes that consistently have twins or triplets do so no matter what pasture they are on at breeding time. Ewes should be in good health, of course, at breeding time, but I think there is more danger from getting them too fat if they get too much good pasture than there is danger of fewer lambs from not flushing. The time that ewes need extra energy or nutrition is during the early part of lactation.

I feed winter hay out on the pastures when the ground is frozen or not too muddy. The sheep waste very little of it despite what manufacturers of hay feeders say. The sheep clean up hay especially well if it is fed on snow. Sometimes a grazier will say that sheep waste hay fed on the ground when the problem is that the hay is not of good quality. That same hay, fed in a manger, is pulled uneaten out onto the floor, where it goes unnoticed in the bedding.

You can rejuvenate areas in the pasture where the grass is thin or the soil poor by feeding hay there. As the sheep congregate to eat, their droppings also congregate, enriching the soil. Some clover and grass seeds in the hay fall to the ground, and the trampling of the sheep works the seed into good contact with the soil surface, where in spring it will sprout and grow. The use of controlled trampling can in fact take the place of a light disking or cultipacking if done when the soil is not too soft or not too hard or frozen. A flock fed on a declining sod will lightly churn the soil surface for better seed germination after broadcasting. The true grazier loves this little trick of adding new forages to a pasture because the sheep do the cultivating and they have no tires, gas tanks, or oil pans to refill. On a pasture of, say, red or ladino clover gone to seed, those biological cultivators will trample some of the seed to the ground as they graze, too, and a whole new crop will spring to life for next year's grazing. The sheep in this case have become planters as well as cultivators and harvesters. This is why historically sheep have been referred to as having "golden hooves."

You don't need a barn to protect sheep in winter. A windbreak will do. One morning when I took hay out to the field for my ewes, it was eleven below zero. They had slept not in the barn, although the open door beckoned, but along the edge of the woods. They romped around as if it were May, waiting for me to get the strings off the bales. Some of them had already been pushing their noses through six inches of snow, looking for old grass underneath.

If you shear in very cold weather, the sheep probably should have a shelter for the first few days. It is nice to have a barn for shearing, too, especially in winter. Shearers do not want to work on a sheep covered with snow. Once a flock grows beyond just a few animals, some kind of handling system for worming, trimming hooves, tail docking, and castration is desirable. Handling systems of gates, chutes, and clamps are a necessity for a large flock. With only forty sheep, we crowd them into a pen in our little barn for all these operations. Sheep are easier to handle when they are crowded together. They feel more secure and can't jump around even if they want to. I equipped one manger with a crude homemade head lock, where I can contain a ewe if I want to milk her or try to force her (unsuccessfully in every case but one) to take back a lamb she has rejected.

A good sheepdog is a treasure beyond price in holding sheep in a fence corner for you to catch or in herding lambs into a truck or into the barn. I don't need a dog because my sheep usually come or go when I ask them to. I always call them when I am putting them into a new paddock, although I don't need to because they crowd around as soon as I approach any gate. But, associating my call with good food, they come running when I call, even if I am standing at the barn and they are in the farthest field.

Guard dogs may help guard against predators, but sometimes they want to protect the sheep from humans venturing into the pasture, too. In talking to shepherds who have tried guard dogs, I get the distinct notion that the practice is iffy and the dogs are not cheap. Some dogs do guard and some dogs don't. The dogs, I'm told, must be raised with the sheep. If you buy one, I'm sure you will get all the advice that's available. For the kind of operation I run, guard dogs aren't necessary. A predator-proof fence is better.

To catch a sheep in the paddock, I bring along some morsels of hay to lure them. They will crowd around me, and I can generally catch the ones I want to examine with one swift grab. Also, when the sheep are feeding on a haystack, they eat holes into the stack far enough that they can't see anyone approaching from behind. Only their butts stick out, a ludicrous sight. I just walk up and grab the one I want. I also have a shepherd's crook, one of the metal kinds that hook a sheep's hind leg. It actually works. I've never been able to catch and hold a sheep with those wooden canes.

Since I can get wood pallets for free, I am always trying to find something new to do with them. I built a kind of stair-step arrangement of them behind the truck for the sheep to walk on when I load them from the barn onto the truck. Animals much prefer going up steps than up a slanted chute.

Occasionally, it is convenient to make a little shelter for a ewe and lamb in the pasture, or a ewe about to lamb. For this I wire two wood pallets together, V-shaped, for a crude windbreak. Another pallet laid on top makes a suitable temporary rain shield. You can imagine all sorts of makeshift variations on that idea, made of plywood or even metal barn roofing.

Some shepherds who lamb in colder weather still do it outdoors, using, when necessary, little shelters like those described. They say that this method runs less risk of pneumonia than lambing in barns that are not well ventilated. I occasionally carry a new lamb into our little barn when it is raining hard, with mother right at my heels. I am probably wasting my time. Healthy lambs that begin nursing right away are very tough little souls. As the old folktale points out, it is generally the shepherd who needs the shelter.

9

Milk and Beef from Pasture Cows

Producing milk profitably on pasture without feeding grain is a very revolutionary idea. Some people, such as the Apples mentioned in chapter 3, are doing it, but they have their own market and can price their milk accordingly. In another interesting development, a new company exhibiting at the Ohio Ecological Food and Farm Association (OEFFA) conference in 2003 offered cheese made from totally grass-fed cows. It was ultradelicious. The company wanted to expand into all-grass milk but had not yet found enough dairy farmers willing to give up grain entirely. Which is to say that they couldn't offer a premium that the farmers figured would make up for the loss in volume. What that really means is that most farmers can't imagine not feeding any corn at all.

Nevertheless, I have a hunch that pure grass-fed milk is being produced and sold somewhere, and I'm just not aware of the particulars. It lacks only enough consumer demand at the moment. We are already seeing what consumer demand has done for total grass-fed beef, a product that seemed unimaginable to the commercial meat market in the United States even ten years ago.

F. W. Owen, a most interesting grass farmer whose views I have become acquainted with from his website (http://www.bright.net/~fwo/index.html), says it all. He was feeding as much as thirty pounds of grain per cow per day in 1994. By

1997 he disagreed with his past views on that subject. Here are his words from his website: "I now think zero grain is the most profitable amount in Ohio. But lots of folks want to participate in the registered business and need milk production records equal to the competition. I say 'let them feed extra grain, it won't hurt much and it will make them happy.'" So I will content myself with writing mostly about making milk with much less grain rather than none at all. I would rather stand up in front of a room full of young Americans and tell them that they don't need cars anymore than tell a room full of dairy farmers they don't need grain anymore.

The easiest way to cut grain feeding is to just take a deep breath and buy a lesser amount of grain than thought to be required, as Nathan Weaver, featured in chapter 3, did. But first make sure your pasture and hay are of high quality, following the methods already outlined and those described in later chapters. Another way is to plant grain, whatever the amount thought necessary, and then graze it, saving the high cost of harvesting and storing it. This practice is just getting started with corn and is considered more an alternative for producing beef than milk. However, there is no reason, other than traditional obstinacy, why corn can't be strip-grazed by milk cows too. Then it will be only a matter of logical progression before cereal grains are strip-grazed, too. Experiments last winter demonstrate that winter grazing of both grain and forage of oats is quite practical.

Many dairy farmers, and I was previously one of them, are convinced that they need grain to get the cows to come into the parlor. But I watched, with some amazement, Mr. Weaver's cows saunter into the parlor without being enticed by grain. In fact, Mr. Weaver said he had to limit the cows' grazing or they would be too full to eat any grain! They got their ten pounds per day after they were milked and left the parlor. They stood patiently, untied, during milking. I was reminded of the milkmaids of yesteryear, who milked their cows untethered out in the pasture. Mr. Weaver's cows rarely defecate during milking, either. He keeps track and talks about going through a week or more of milking thirty cows without even one poop in the parlor.

Corn has added appeal for winter grazing because the ears almost always stand above the deepest snows. But dairy farmers are as hesitant about year-round grazing in the North as they are about giving up heavy grain feeding. They believe that milk production would drop if cows had to go out in very cold weather to graze for their grain. But surely they could do so for part of most days, during which they could get the supplemental grain energy they need to maintain production.

Another interesting objection that some dairy farmers, especially the Amish, have to year-round grazing is that there would not be as much barn manure to spread on the fields. I would think that they would rejoice over that. But no. "We believe that manure and bedding building up together in the barn and partially composted has more fertility value than fresh droppings scattered over the pasture," says David Kline, the editor of *Farming* magazine in Holmes County, Ohio, who has extended his grazing season to nearly nine months but keeps his cows sheltered in the barn in the coldest months. There is no doubt that composted manure has more nutritional value than fresh, or at least the nutrients in compost will not leach away as fast as those in fresh manure when spread on the soil surface. Enough nitrogen is lost in freshly deposited droppings to justify the cost of hauling barn manure in some cases. But my sheep have their preferred places to bed down in every paddock and these places over the course of a year get what I imagine is the equivalent of the amount of manure/bedding that farmers apply per acre with the spreader. I don't see how grass could be made to grow any better than mine with those fresh droppings and urine. If the grass grew any better I would need a bulldozer to get through it.

Oxen can figure in specialized kinds of pasture farming. When garden farmers start thinking outside the box, or rather, outside the stanchion, novel things start happening. I recently learned of a young man, age thirteen, who is training and using oxen for draft power on his parents' small farm. Imagine in this day and age a boy who for Christmas asked for money to take courses in how to become a drover—and got it. His parents took him to Tiller's International, a long-running school devoted to teaching traditional rural skills (10515 East OP Ave., Scotts, MI 49088). Now he uses his pair of milking shorthorn steers to work around the farm and in historical reenactments in the area.

Oxen fit into pasture farming as well as horses do because there is no need to feed them heavily to fatten them for market or for super milk yields. They can, nevertheless, return the investment in them by providing meat when their working days are over. You can't make oxtail soup out of old tractors. The popularity of draft cows in third world farming stems partly from the fact that the cows give a little milk for the family in addition to providing draft power. You can't do that with a tractor, either.

Good Meat on Grass

Meat produced on pasture without any grain is more common than no-grain milk. But the most preferred method is to finish the beef with a little grain toward the end of the pasturing period. Total grass-fed beef is gaining adherents as the news goes around about the possible problems in grain-fed and hormone-treated beef. Demand, at least in northern Ohio, is outpacing supply. Articles declaring that grass-fed meat tastes better than grain-fed meat appear regularly in urban and consumer magazines. Even cookbooks are advocating grass-fed beef and showing how to fix it to get rid of any toughness or stringiness that the corn monopoly claims characterizes grass-fed beef. As Carson Geld, an American who became a Brazilian rancher and farmer years ago, tells me: "Brazil has been eating and enjoying good beef without corn for centuries. We haven't had corn. We don't even have a grading system like the U.S. does. Our beef is much less expensive. We couldn't afford to feed corn if we had it. Those who think the meat is a little tough use papaya juice to tenderize it."

One good way to produce grass-fed beef that is just as tender and juicy as corn-fed is what is called baby beef. The cow is bred to calve at the start of the lush grass season. Mother and calf have access to good pasture at all times. The calf nurses up until the day it is butchered, at about 650 pounds and 9 months of age, as winter approaches. What is required is a mother cow that gives more milk than a regular beef cow but less than a typical dairy cow. Angus/Jersey and Hereford/Guernsey crosses are the ones I have used, but others would surely do as well.

A six-hundred-pound calf nursing a twelve-hundred-pound cow sounds dangerous to the cow's udder, but we have never experienced trouble. The calf by that age is in the process of being weaned naturally, and the cow is no more affected by its butting than a ewe with a lamb that is being naturally weaned. In fact, the calf may not be getting much milk from the seventh to ninth month of its feeding period. It has, however, been getting more milk in the first six months of its life than beef calves usually get, and also better pasture. That is why the meat is so tender and juicy. It is no longer veal. It is a little firmer than veal, pink in color, and much more flavorful than veal. Nor is it fat in the way grain-finished beef is fat. Fat lends flavor to meat, as everyone knows, but different kinds of fat lend different kinds of flavor. The flavor of grass/milk fat is less cloying and does not dull the appetite like the flavor of corn/soybean fat. And when you eat baby beef, at least our baby beef, you don't have to worry about hormones and antibiotics. Baby beef can be raised easily without hormones. We have never had to use antibiotics since we quit weaning calves. Calves that are not weaned at a young age rarely get scours. Commercial dairy calves so often need antibiotics because they are weaned in three days or less.

One of baby beef's advantages for garden farmers is that the mother cow can provide milk for human consumption while she is nursing the calf. Our Guernsey/Hereford cross gave more milk than the calf could drink in the first two to three months. During that time I milked her once a day. She soon learned to hold back her milk from me, so I had to milk two teats while the calf was nursing the other two. Not even a cow is smart enough to hold up her milk in two quarters and not in the other two. Some garden farmers keep a dairy cow that gives lots of milk so that they can draw off milk for the table during the whole lactation period, or so that they can put two or three calves on the cow. Getting a mother cow to accept other calves is easy with some cows and difficult with others. Once a cow gets used to the idea of adopting calves, she will usually cooperate more readily the next time.

The disadvantage of baby beef comes when you try to sell a 650-pound calf on the corn-ruled market. It will not grade high enough to command a good price because it is not marbled the way corn-fed beef is. Also, it is too easy to cheat. A beef buyer can tell a steer finished properly on grain just by looking at it. And there are ways to measure carcass quality and fat when the definition of quality is grain-fed. But methods to grade the quality of baby beef before it is butchered are not much in evidence. A 600-pound calf that has not been on a good milk/grass diet may look just about the same as one that has. And just because it is "organically certified" does not mean that it is necessarily tender, either.

To compete in the commercial market, producers of baby beef need to establish their own customer lists and team up with local butcher shops. This is already happening, not only for baby beef but all beef. In fact, the business is exploding. The two local butcher shops I am familiar with have so much business that they have to turn people away. A cousin of mine operates one of them, and he says the number of people wanting to avoid hormones and antibiotics in the meat they eat is increasing. Such people buy a whole or half carcass directly from farmers who cater to this trade. These customers often prefer the smaller-sized baby beef because they can't handle a large amount of meat in their freezers all at once. With regular-sized steers, two or three customers will divide a carcass, and the local butcher shop will cut and package the meat accordingly.

Attempts to develop smaller breeds of beef are ongoing, partly because of the demand for "freezer" beef and partly because smaller-sized cattle work better on small pasture farms. The danger of too much trampling in wet weather is less. We recently visited a farm in central Ohio that raises and sells miniature Herefords. Miniature Jerseys are becoming available, too. The increasing popularity of Dexter cattle, a small dual-purpose cow (meat and dairy cow) is another example. Dexters were developed in Ireland, where garden farms have long been a part of the social fabric.

The miniature breeds are quite expensive because they are rare. I can't afford one. It might die. So I'm going the poor-man route. I recently purchased a heifer calf from a strain of Hereford more like the breed was fifty years ago before efforts to develop larger-framed beef animals started. My thinking is to breed my calf, when she gets old enough, to either a Dexter or a Jersey of the smaller type common before the great desire for large size in all livestock gripped the husbandman's mind. If I live long enough, I hope to arrive at a cow that will be just right for the garden farmer without being dwarfy, as some miniatures are.

At any rate, cattlemen fairly well agree that the smaller beef breeds are more adaptable to finishing on grass alone than are the bigger breeds. "Grass Finishing Beef," an article by Henry Bartholomew and Fred Martz in the July 1995 *Stockman Grass Farmer*, put it this way: "This system will not work for large frame animals. Those big guys will just keep getting taller and taller and taller and . . . bigger. It will work best for producers with small to medium frame cattle that will finish at 1050 to 1200 lbs." As an indication of how far away the United States is from total grass finishing, this article spent about half its space talking not about truly grass-finished beef, but how much the grain ration could be decreased while still having meat that tastes grain-fed.

Beef animals have to be selected that will respond well to grass finishing just as we have bred animals that respond to corn finishing. It takes time. It takes commitment. In fact, just as I write this paragraph, on February 28, 2003, there's a conference going on in Kansas City on this very subject. (You can reach the people involved by contacting *Acres U.S.A.,* another publication that has long preached grass farming, at http://www.acresusa.com.) The North Devon and the British White are two somewhat exotic breeds that some breeders propose as a good place to start.

I used to get excited about exotic breeds, but I'm too old a dog and have seen too many exotic breeds come and go. They might be exactly what is needed to develop pastured beef cattle, or they might not. My bias is that Hereford, Angus, and Jersey crosses will get you there just as quick and much cheaper. One good source of information and stock in this regard is Keeney Angus, 5893 Hwy 80 W, Nancy, KY 42544. I have never met Mr. Keeney, but we talk on the phone. He is one smart cattleman.

Culture is a powerful influence when it is also backed by economic reality. Before we can produce grass-finished beef we have to think like a New Zealander or Irishman or cowboy on the Pampas where grain is either not available or too costly. If our thinking is merely to use pasture as a prelude to finishing beef on corn, the notion of grass-finished beef in America will never quite happen. In the face of cultural resistance, the transition must take place by increments. If the market demands corn-fed beef, then graziers will first start thinking of corn as a pasture crop, not a harvested grain. Pastured corn could satisfy the present demand for corn-fed beef but still save huge amounts of money and labor.

I think of an experience in my teen years that taught me a good lesson about how slowly cultural and agricultural change comes. Our neighbor Fred Schmidt was an excellent farmer. On merely eighty acres of good soil he made a comfortable living. My father rented the farm when Fred retired, so I was often in his company. While we worked together, he would regale me with his farming experiences, some of them over and over again. He found particular pleasure in retelling the time he "made a killing." One year in the early 1940s, he decided that cattle prices were going to be high that winter. He did something unusual for his cautious, conservative nature. He bought a train carload of western steers to feed out over the winter. Before he bought the steers he decided on another course of action even more unusual. He had already cut and shocked his corn crop, and he was not particularly looking forward to husking all that corn, cribbing it, and then grinding it into meal to feed the steers. What if he simply hauled the shocks to the cattle yard and fed them by the bundle, ears of corn, fodder, and all. Why not? And so he did, hauling in a shock or two every day and scattering the bundles on the ground for the steers to eat. They fattened fine, the market did go up, and he "made a killing." "The point, Gene," he said to me, pointing his finger, "is that I didn't have to do all that work of husking, cribbing, and grinding the corn. The steers did it."

I understood his satisfaction, but the reason I never forgot the story (other than that he told it to me at least a dozen times) was that he could have saved a whole lot more labor. Since all his fields were properly fenced, he could easily have turned the steers into the standing corn as we were doing with hogs back then, and they would have done the whole harvesting job by grazing. He wouldn't have had to cut and shock the corn. He wouldn't have had to do *anything* except watch the steers get fat.

I used to think that cattle would bloat if turned into a cornfield, especially if the corn ears were still not completely dry. Now that graziers are strip-grazing corn, we know that fear is ungrounded if the animals are introduced to it gradually. Moreover, I was able on two occasions to observe where steers actually roamed cornfields for days. Kind of amusing, actually. In one case, it took the owner, a laid back kind of guy, several weeks before he noticed his beef cows were entering and exiting the cornfield through a hole in the fence and were feasting on the corn. I knew, but I said nothing for awhile. I wanted to see what would happen. No bloat. No problem really except that the cows ate some of the corn and knocked over a few stalks. They were very happy cows, as a matter of fact. The other case, where the farmer was even more laid back, a steer got out into his tall corn in August and sort of went wild. Have you ever tried chasing a steer through hundreds of acres of unfenced corn? The farmer grew wise. As long as he didn't chase, the steer didn't trample; instead, it sauntered lazily down the rows, jerking off an ear whenever it felt the urge. The farmer decided that he'd just let it go till harvest and then shoot and butcher it. So the steer spent the summer in the tall corn, eating its fill. By fall it was so fat that it didn't like to run anymore, and it strolled back to the barn fat and sleek and ready for market.

Some farmers fear that corn will be wasted if grazed. The cattle will trample more than they eat. But where corn is strip-grazed, the cattle clean it all up. Raising hogs with the cattle would be a way to make sure there was little waste. Besides, why all this preciousness? What about the corn that is wasted in fields that are mechanically harvested? In some years many ears drop off from drought or some weakness in the corn variety. There is no way to mechanically pick up those ears. In a pasture system, cattle, sheep, or horses can clean it up.

And in fields mechanically harvested, the fodder is usually "wasted" that is not grazed, as it used to be, by workhorses and dry mother cows (actually nothing is wasted in nature—that fodder makes more organic matter in soil). Cost-conscious farmers bale up the fodder for feed. Again, the old habit. Take expensive machines to the field to harvest something that the cows could harvest for themselves.

One of the best essays about grazing cattle that I have read comes from Jim Gerrish, a leading pasture scientist at the University of Missouri Forage Systems Research Center in Linneus, Missouri, who has now resigned to begin a different life. (I found the article in Ohio State's *Grazing Newsletter*, but Mr. Gerrish has made most of the same observations in various articles in the *Stockman Grass Farmer*.) The essay sums up what Mr. Gerrish has learned in twenty years of experimenting with grass farming. Basically, he makes the point that grass farming is about intensifying management of space or landscape over extended periods of time, not about grazing itself. And the main management goal is not to increase beef production per acre but to lower production costs. So true on all accounts. As merely one example, and not the most important one, here's what he says on the subject of corn: "Then we started exploring other alternatives for winter feed for grazing animals. The crackpots were showing up again and they were grazing standing corn, of all things. If we were going to feed corn to cattle, why not let cows do the harvesting rather than a $120,000 combine. You can buy 200 cows for the same price as the combine and cows can have babies and combines never could. We were beginning to think like grass farmers. The revolution had begun." He is using the word *crackpot* in an admiring, not derogatory, way. What amazes me most about this passage is that I found it only *after* I had written the paragraph about Fred Schmidt above. Mr. Gerrish and I had arrived at the same "outlandish" notion independently. In fact, we have never met. I was, and still am, one of his crackpots—and I proudly accept the label.

If the cattle and dairy industry can begin thinking like complete graziers, they will go on from the concept of grazing standing corn to grazing standing cereal grains for the extra supplemental energy and protein they might need to produce for the market. Then the revolution will be complete, because we can grow cereal grains (oats, wheat, barley, rye, triticale, spelt) just like we can grow pasture grasses and clovers—that is, with little or no cultivation, expensive planters, or expensive harvesters. We can frost-seed or fall-seed these grains by broadcasting into existing but declining stands of clover or on land well trampled by grazing animals. It works, and I know it works because I've done it.

One advantage of grass-finished beef, seldom mentioned, is that it encourages raising beef from calf to finished meat all on the same farm. This encouragement would not be welcomed by the beef industry as it now operates. Currently, a calf spends the first part of its life running the range with its mother. It is then weaned and shipped somewhere else to go through several stages as a "stocker" before it finally ends up in a feedlot, where it is fattened and finished and shipped to the packer. A saying in the beef industry is that the average beef steer has seen more of the United States than the farmer who fattens it for market. Obviously, a whole lot of inefficiency creeps into this process that is supposed to be very specialized and efficient. Think of just the huge transportation costs involved. And all that transportation means stress on the animals, which means more sickness and death loss. But because a cow-calf herd can be profitable on rangeland that would not put the necessary weight on stockers, which in turn can be grazed on land where corn for finishing the feedlot beef would not be profitable, maintaining the three stages of beef production in three different environments seems efficient. Sometimes it is. But the three-stage system is far more complicated and incomprehensible than that. More often the calves, the stockers, and the corn are shipped to huge feedlots where the profit, if any, comes more from tax shelters than from efficiency. The entire beef business exists on wheels turned by fuel-gulping piston engines. If written today, the old song about "home, home on the range" would be "home, home on the highway."

Compare that grand travelogue with a pasture farm where a calf, or at least a stocker, can spend its entire life and be sold locally or in nearby cities. The fattening beeves go into a paddock first to eat the best of the grasses and legumes. When the fattening cattle move on to another lush paddock, less-demanding livestock, like dry cows and recently weaned calves or stockers, follow up. Think of semi trucks hauling stockers to feedlots while tearing up the road and causing thousands of accidents. Now think of the stockers walking from one paddock to the next.

10

The Rising Farm Interest in Goats

I have only a little experience with goats, and so I don't think I should pose as being expert enough to write about them. But I must. Goats as producers of both dairy and meat products fit better in pasture farming, especially as a garden farm project, than cows or sheep. Perhaps because of that, but more likely because of the steady influx of new Americans accustomed to goat milk and goat meat, the interest in goat husbandry is increasing. Since I am putting myself far out on a limb by predicting a future in which much more of our animal food products will be produced in small, even backyard, pasture systems, I must, logically, also predict that the goat will be at the forefront of this new exurban society. Already goat milk is consumed in more areas of the world than cow's milk. The reason is that goats are easier to manage in low-cost, garden-farming arrangements.

Goats are better than cows for the family striving to produce all its own food on a small acreage. Here's why:

1. An acre or less can furnish the hay and grazing for a doe and two kids.

2. Goats are a whole lot easier to transport than cows. Many goat owners with limited space take their does to a goat farm where bucks are kept for breeding. That's cheaper than keeping a buck and avoids the possible problem of buck odor. I've seen goats riding in the backseats of station wagons.

3. A good doe provides as much milk as a family generally needs. The milk is more digestible than cow's milk and is in fact the only milk for lactose intolerant people. Some families keep a goat because they have children allergic to cow's milk.

4. Goats are triple-threaters. Some are dairy goats, some are meat goats, and some are both. And some, like Angoras, provide wool for weaving.

5. And, at the bottom line, goats make amusing pets. Cows do, too, sometimes, but you can hardly let a cow follow you around the yard all afternoon.

I once dined with my homesteading heroes, Harlan and Anna Hubbard, now passed away. Harlan and Anna lived along the Ohio River as independently as it is possible to live in the world today. They produced all their own food or got it from the wild. They had built their own home. They eschewed electricity and automobiles. They had only a small income. But they lived comfortably, even grandly, to my way of thinking. Their goats provided them with almost all their meat and milk. The goats ran mostly in the woods, browsing their food and costing the couple almost nothing. I drank the milk, kept chilled in the Hubbard root cellar. I ate chevon that the Hubbards had canned. The milk was as good and sweet as any I've ever enjoyed, and the creamed, shredded chevon was delicious. There was no grain in the goats' diets except perhaps some leftover sweet corn from the garden.

At the other end of the goat-keeping spectrum are ultramodern farm enterprises whose aim is to make a good profit from selling goat milk, goat cheeses, and fancy gourmet foods made from goat while acquainting people with what the animals are really like. For example, as anyone should be able to deduce, goats don't eat tin cans. (They might, however, chew the sparkplug wires on your tractor, just as cows will do.) These sophisticated and often ritzy farms are becoming as commonplace as the many local wineries popping up, and they are similar to them in the way they accentuate quality and assiduously cultivate the public. The most startling example I know about is Caprine Estates/Willow Run Dairy, about fifteen miles southwest of Dayton. The goats there live better than many people do. In residence are more than a thousand goats of the six best-known breeds: Nubians, LaManchas, Saanens, Alpines, Oberhaslis, and Toggenburgs. The goats are milked in a state-of-the-art milking parlor. Computers keep tabs on all aspects of their behavior. Even the amount of grain they eat is regulated electronically.

Caprine is not exclusively a pasture farm in that it does feed grain, but I am sure the opportunity for selling all-grass (no-grain) goat milk or meat would be very good because of the interest in all-grass cow products. (The only book I know about that deals with goats specifically from the pasture farming point of view is Sylvia Tomlinson's *The Meat Goats of Caston Creek,* available from Redbud Publishing.) Pasture farming with goats would be comparatively easy without grain because quantity is not nearly as strongly emphasized in goat dairying as it is in cow dairying. There would not be as much temptation to provide a heavy grain ration to increase production. The forage plants used would be the same. The best goat meat is cabrito, which is meat from milk-fed goats butchered at about eight weeks of age before feeding grain would have much effect on it. Chevon, meat from goats of about six to nine months in age producing about twenty pounds of meat, should be as high in taste and tenderness as lambs on good pasture. Goat meat is slightly higher in cholesterol than beef or chicken. It has fewer calories than beef and about the same as chicken. The taste is not like other domestic meat, but has its own unique flavor. I find it more like beef than lamb. I am told that the secret to good-tasting chevon is to cook it slowly to preserve moistness. But that is true of most meats, in my experience, especially venison.

Yes, goats are by nature browsers not grazers, but most of the modern breeds graze grass as thriftily as cows do. Goats will eat seedling trees and clean up brush and thorny multiflora rose better than cows or sheep. Many manuals of instruction point out that goats can eke out a living on rough terrain where cows would not be economical, which is why goats are favored in poorer, third world countries. But for practical production of meat or milk, pasture farmers will want to provide a quality pasture for goats, the same as they would for sheep or cows.

Goats are harder to keep fenced in. An old goatherd saying states that "if your fence won't hold water, it won't hold goats." That's one more reason for installing woven wire or panel gate fencing.

Dairy or meat goats will thrive on a pasture system that sheep and cows thrive on, which is to say, a pasture of mixed legumes and grasses. Goats should be ideal for grazing cereal grains going to head because they like to browse on plants a few feet off the ground, whereas cows or sheep prefer to graze with their noses down on the surface of the field.

Another reason why goats work well with pasture systems is cultural, and I mean human culture in this instance. People who decide to raise goats usually do not come out of traditional agriculture and so do not have the usual prejudices against pasture farming or goats. They know that, properly handled, goat milk does not taste goaty. They know that male goats can have a rank odor sometimes, but it smells no worse than pig manure. If you want to smell something *really* awful, pen a cat and force it to live standing in its manure and urine-soaked bedding the way buck goats are often kept.

Also, the goatherd, not coming from a traditional farm background, has no preconceived notions about grass farming. The idea of dividing up a backyard into sections and growing good grass and clover on it for a goat instead of for the lawn mower seems entirely logical and practical.

I once milked a goat for the specific purpose of tasting the milk. It tasted just like cow's milk to me. Milk from any animal picks up tastes easily. Unpasteurized milk from cows on fresh pasture has a stronger taste than unpasteurized milk from cows on good hay. But you soon get used to the change from winter hay to spring grass. If goat milk has a "goaty" taste, the reason is poor quality feed or, more likely, careless handling of the milk. Beginning backyarders are not likely to appreciate the fact that you can't leave any milk unrefrigerated even for a short length of time. The quicker you get it cooled down to thirty-eight to forty degrees, the better it will taste.

Hispanics who were brought up on chevon expect it to taste like chevon. People reared in a culture in which goats are often used as symbols of the devil might approach chevon with suspicion. They expect it to taste like lamb or pork or beef, and when it doesn't, they are disappointed. I grew up despising the taste of lamb. That's because I never tasted real lamb until I was older. What we ate, or tried to eat, was old ewe—something Dad could not sell for a good price. Nor did Mom know how to prepare mutton the way a good restaurant prepares a rack of lamb. The first time I tasted lamb in a restaurant I was amazed. It was delicious and there was not a trace of that "old ewe" ("goaty") taste that I had known at home and at boarding school.

Goats like to climb up on anything they can. It comes from their heritage of jumping around on rocky mountainsides, I suppose. If you try to let them self-feed from a pasture haystack, they will climb on the stack, wear it down, and waste more hay than they eat. Sheep will sometimes do that, too. If you have low buildings, expect to find your goat on the roof occasionally.

There is only one real problem with goats. They are so lovable that you might not be able to kill and eat them. My favorite goat story was told to me by a family homesteading in the fastness of Appalachia in eastern Kentucky. They kept goats. One day a quick mountain storm passed over them. The children were playing in the woods and couldn't make it back to the house before the rain hit. They sought shelter in the outhouse. The four goats piled in with them.

11

Root, Hog, or Die

Pigs are remarkably adaptable animals, which is why they have for so long been an integral part of farm life, the "mortgage lifter" of traditional agriculture. They will adapt to almost any agricultural scheme, however whacky. Hogs were popular with pioneers because they could thrive on acorns and roots, which led to the folk saying "root, hog, or die." Even today hogs escape to the wild and root for a living. They are so genetically congenial that clever humans have in less than half a century changed their original lumpy, dumpy shape into carcasses as long as school buses and just as lacking in fat. They will survive in crowded animal factory cages. They will live contentedly on human food garbage if forced to do so. In fact, so well will they respond to human manipulation that we now have hogs that can live *only* in animal factories. If they are let out into the cruel weather of the real world, they are apt to die of pneumonia.

The modern hog is a softie; never mind that its ancestors thrived in the cold winters of northern Ohio for generations. Farmers without the latest hog factory facilities, but who nevertheless insist on sows gestating in the dead of winter, think they have to use heat lamps to keep new pigs warm. Eventually, they use the heat lamps even when it isn't so cold. But, unlike really dedicated husbandmen, they don't necessarily stay with the sows night and day when they are birthing. The sows root their bedding up against the lamps. The bedding catches on fire. Good-bye barn, good-bye pigs. It happened on our farm when I was a kid, but fortunately Dad spotted the smoldering straw and put the fire out before calamity struck. It happened again in our community just this week. The barn and all the pigs in it burned up. The owner was at work on a night shift at a local factory.

This is the tragedy of trying to make biology behave like business economics. First of all, if pigs are such softies, we should not be farrowing in the depth of winter. Second, the sows involved, at least in our operation sixty years ago, would have birthed their pigs in their nice strawy pens without need of heat lamps at all. It was the utility companies, intent on finding more ways to market electricity, that convinced farmers to use heat lamps. I wonder sometimes if the sow was not heaping straw up against the lamp to block some of that excess heat.

Crushing, the word used to refer to the way sows inadvertently lie on their baby pigs and kill them, was a common problem of the pigpen before the farrowing crate. It was caused almost entirely by penning up sows in close quarters instead of allowing them to farrow on pasture. In a pen, it was easy enough for a little pig to get between the sow's huge body and the pen walls. Crushing was also exacerbated by nervous husbandmen making the sow nervous, too. In hog factories, crushing at least is normally avoided. The poor sow is crammed into a farrowing crate, where she can't roll onto newborn pigs. She can't even turn around.

My father would throw chunks of sod from the permanent pasture into his pens of pigs and sows as a source of iron and other trace minerals. He often directed me to do this task, a good job for a boy. I would pull chunks of sod from the top of steep creek banks. Think of what we were doing. Instead of taking the pigs to the field, we were trying to bring the field to the pigs. Strange are the ways of humans.

Some pigs develop sharp "black" teeth, which good husbandmen clip off. This is not done to prevent injury to the sow's teats, as many people think. A nursing baby pig sometimes tries to keep other pigs away from the teat it is sucking on. Throwing their snouts from side to side, they slash each other with these sharp little teeth, which are really tiny tusks. The damage done can be serious. I could not remember my father ever cutting pigs' teeth, so I asked my cousin, who raised hogs for fifty years, why it is now common practice. "I think it is the way hogs are being bred to endure confinement," he said, without hesitation. "We didn't have to do it in the old days. I remember when it started, and even then I only had to cut the teeth on a few pigs. The same thing with tail-biting. I never had a problem with tail-biting until I started raising the modern, lean, meat-type hogs. They are very high-strung and mean to each other."

I leave the reader to his or her own conclusions.

In the 1950s I was raising pigs in Minnesota, where the temperatures easily dropped to fifteen degrees below zero Fahrenheit. We did not use heat lamps. Minnesotans had been raising pigs from at least the 1880s until the 1940s without a spark of artificial heat, electric or otherwise. They had enough brains not to schedule farrowing during the depth of winter, no matter what need there seemed to be for cash flow. In fact, "cash flow" was not a term common in farm language then. Sows were farrowed in fall and spring. For both sows and growing pigs we built crude winter shelters of saplings, board gates, or two-by-fours and then covered them completely with corn fodder, straw, or haystacks. Those bare-bottomed piglets snuggled down inside these cavelike shelters in January as warm as a nudist on a Caribbean beach in June.

The Minnesota farmer I worked for also taught me the advantages of pasturing hogs on alfalfa pasture. He turned sows and pigs into alfalfa so tall the pigs disappeared from sight. He fed only about half the corn that hog producers feed today. For years I thought that idea was original to him. But I have since then found a profile of a hog farmer in the old, defunct *Farm Quarterly.* In the 1940s he was grazing his hogs on alfalfa *year round* in Missouri. I consider the article, titled "Ton Litter Production" (in the 1950 autumn issue) to be one of the most important pieces of evidence from history on the practicality of pasturing pigs. J. Ward Stevenson was the farmer's name, and I wish I could have talked to him, mainly because the article did not give the details that a pasture farmer craves. Mr. Stevenson ran his sows, pigs, and shoats continuously on alfalfa pastures, starting two to five days after the pigs were born. In Missouri, he said, the hogs could find alfalfa to graze all year except after heavy snows, if the pastures were rotated properly. The article unfortunately did not say how much grain he did not have to feed because of this kind of pasturing. But it is interesting to note that 'way back then, Mr. Stevenson anticipated what would be happening today in pasture farming on marginal land. He had sold off part of his land to make his operation more efficient and had given up corn farming altogether. He believed it was cheaper to buy corn and spend his time raising litters of pigs that weighed in at a ton by market day. It was cheaper, of course, because his year-round alfalfa pastures were saving considerably on the amount of grain he would otherwise have had to feed.

Another interesting detail of Mr. Stevenson's method of raising hogs was that he could so arrange his pasture plots by using movable pens that he would turn a sow and her pigs out as a matter of course, on pasture that had been free of hogs for the previous two years. By using clean land this way, he avoided internal parasite problems that are as endemic to hogs as they are to sheep. His stocking rate was two sows and their litters per acre.

The pig's adaptability to almost any environment is the main argument against the hog industry's claim that the only profitable way to produce pork is in its factories. The industry has done another of those self-fulfilling prophecy tricks. It encouraged a type of hog and a type of economy that fit its production methods whether the hog liked it or not. Then they claimed the corn and cage method was the only way to make money with hogs. The result has been that the hog's status changed from mortgage lifter to mortgage keeper. The more factories that jump into the business, and the bigger they get in a vain effort to gain steady profitability, the more the hog cycle goes up and down, from short boom to long bust.

Imagine that you are a pioneer farmer in Illinois raising hogs in 1840. You let your hogs run in the woods most of the time. Root, hog, or die. You feed your pigs some corn, if you have some. More likely, you feed them some potatoes and turnips. If you are real clever, you pen them in a field, where they will root for grubs and plant tubers, providing the primary cultivation for next year's crops. All you have to do after that is drag a small cedar tree over the rooted-up soil to level it a little and plant. (The ancient Anglo Saxon words for pig and plow come from the same word, which meant, literally, "to root in the ground.") When you want to market your hogs, you drive them to the canal and maybe even stitch shut the eyes of the orneriest ones with hemp string to make them drive easier and lead the rest on. At a canal dock, you load your hogs on a boat and ship them to Chicago. Then you count your money, not much, but almost all profit, buy your wife a new dress, go home, fry yourself a slice of ham, and wash it down with a bottle of Chicago's choicest rotgut. Ahhh, the good life.

Then one year, along comes this scarecrow missionary, like John the Baptist crying in the desert. After he cons you out of a good meal, like all missionaries know how to do, he shakes his head sadly and tells you that you aren't raising hogs the right way, you dumb farmer. Someday, he says, wagging a finger, farmers will put hogs in buildings big as churches, even heated and air-conditioned (you have no idea what air-conditioned means, of course). Farmers will put their sows in cages on concrete floors, says the missionary (you don't know what a concrete floor is, either), and hogs will never wallow in the mud. They will drink their water out of spigots, and by switching a button with their noses, a mixture of milled corn and soybean meal (you have no idea what soybean meal is—there wasn't a soybean in the United States until about a half century later) drops into their troughs. A fat hog will in those days, the missionary predicts, reach the end of its days having never stepped on the earth or seen the sky.

How would you react? You would tell the guy that he was full of hog manure, only you wouldn't use the word *manure*, and then you would usher him out the door with a suggestion that, if he ever came back, the sheriff would be waiting to conduct him to the nearest loony bin. The lesson of course is not to be too quick to usher missionaries out the door. Just as the farmer of the 1800s had no idea how farmers would raise pork a hundred years from then, so we don't know today how farmers will be raising hogs a hundred years from now. We don't even know how we'll do it twenty years from now. We might market most of them as sixty-pound weanling roasters, in which case the pigs might not only never see earth and sky but also never corn.

It has only been in the past fifty years that pigs came in out of the weather to live life in concrete condos. Even when they were being fattened, on corn, of course, they had the run of the pasture. They loved to eat snakes and minnows in the creek when the water got low enough to make catching them easy. They loved to root for worms and plant tubers. If rings weren't put in their noses, they would tear up big sections of the pasture as they rooted, a habit that farmers found deplorable. But there were times when that behavior could be beneficial. Hogs love bindweed rhizomes, and bindweed (also known as wild morning glory or weed-from-hell, or, to my mother, bitchweed) is a curse of modern farming. So the "crackpots" of agriculture have been known to turn hogs into fields hopelessly overcome with bitchweed (true to its nature, this weed is showing signs of immunity to herbicides) to clean up the mess. But bitchweed, again true to its nature, loves to be rooted; the more you disturb the soil, the faster it will grow. Not even the crackpots have yet concluded that maybe the way to raise hogs is on bitchweed.

Instead farmers decided to put rings in pigs' noses to keep them from rooting. The ringed pigs remind me of today's hip, cool teenagers (or vice versa, take your choice). One of the biggest selling points of confinement hog facilities was that farmers no longer had to ring the pigs. Ringing pigs is considered cruelty to animals; ringing teenagers is considered modish lifestyle. Go figure.

A practice much in vogue fifty years ago was "hogging off" the corn—that is, turning them into a field of mature corn and letting them graze it. If the hogs rooted up the soil, it didn't matter because the field was going to be plowed or disked for the next crop anyway. The idea of hogging off the corn was popular because it avoided the expense and travail of harvesting. We tried it. It was my job to cut down a strip of corn every day with a corn knife until the pigs learned they could knock the stalks over themselves. I loved that job; I was Prince Valiant whacking down invading hordes of barbarians. Hogging down corn was a wonderfully cheap and labor-saving way to harvest corn and fatten hogs, but we quit doing it. Grandfather Rall said it was too wasteful, and since he still owned the farm, well . . . .

Hogging off corn is not really wasteful, but we were not practicing a complete grazing system with other animals, too, nor were we strip-grazing the corn with electric fence. In any event, the few ears wasted amounted to far less cost than the labor and expense of mechanical harvesting. But Grandfather Rall was the type who, if he hit a nail off center and it flew into outer space, would stop and hunt until he found it. Time was not of the essence to him because labor (mostly my father's) was dirt cheap. Grandfather Logsdon was another story. He first ran lambs through his corn in August to eat off the lower leaves and the weeds growing between the rows. Then as the corn matured, he turned in the hogs to do their harvest. Then in went his cows to eat the fodder and any ears that remained. Finally, the draft horses and dry ewes and cows wintered over on whatever was left. Grandfather Logsdon understood that life is short and there is no use wasting it doing unnecessary work. He was a grazier and didn't know it.

Many years later, when I thought hogging off corn had gone the way of Egyptian mummies, I ran across an article by Wendell Berry in the *Draft Horse Journal* of spring 1986. The article was about a farmer, Lancie Clippinger, who lived scarcely forty miles from me. Subsequently, I visited the farm myself, but I will let Mr. Berry tell the story in his words:

> Lancie had planted forty acres of corn that year [1982]. And he had bred forty gilts so that their pigs would be ready to feed when the corn would be ripe. The gilts which he had raised the year before, produced 360 pigs, an average of nine per head. When the corn was ready for harvest, Lancie divided off a strip of the field with an electric fence and turned in the 360 shoats and the forty sows. After the shoats had fed on that strip awhile, Lancie opened a new strip for them. The strip where they had previously fed he then went over with the corn picker. In that way he fattened his shoats and sows and also harvested all the corn he needed for his other stock.
>
> The shoats brought $40,000. Lancie's expenses, he told us, were for seed corn, 275 pounds of fertilizer per acre, and one quart of herbicide. He did not say what his total costs amounted to but it was clear enough that his net income from the forty acres of corn had been high in a year when the corn itself would have brought perhaps two dollars a bushel.
>
> And at the end of the story, I remember there was a conversation between Lancie and Maury [Telleen, editor of the *Draft Horse Journal*—Lancie did most of his farming with horses] that went about like this:
>
> "Do you farrow your sows in a farrowing house?"
>
> "No."
>
> "Oh, you did it in huts, then?"
>
> "No. I have a field I turn them out in. It has plenty of shade and water. And I see them every day."

Lancie Clippinger would not have called himself a grazier, but that is what he was. If all the farms in America were like his, and all the farmers thought like Lancie, there would be no farm problem.

Hogs can make better use of corn than ruminant animals, but that does not mean that corn is absolutely necessary for pork production.

Hogs have been traditionally fed on potatoes in Scotland and on barley in Idaho. Around cities, zillions of hogs have been fattened on industrial food wastes and table scraps. That practice still goes on, but now the food garbage must be steam-sterilized to kill possible pathogens, not that the pigs give a damn. The pork so fattened may be a little "soft" and may bring a lower price because it is not corn-fed, not that the people eating it give a damn, either. In taste not many people could tell a difference. We feed all our table scraps to the two hogs we raise every year for our own meat. They also eat corn off the cob (not shelled or ground into meal) and a good helping of high-quality legume hay every day. They get a smidgen of commercial feed, too, my way of guaranteeing to myself that they get the minerals they need, since I do not raise them on pasture. Pasture hogs are in our future, but it may be my son doing it. I have to stop someplace.

If you run hogs on pasture with gestating sheep or cows, be aware that porkers can get mighty fond of the taste of blood. Not only have they been known to eat humans, especially children who fall into their pens, but they will sometimes eat a lamb or calf in the act of being born. Once they start, a pack of them, like a pack of wolves, will even tie into the mother if she stays down. That consideration aside, hogs are rather easy to control with a strand of electric fence about a foot off the ground. Hogs hate to get shocked.

With electric fence as a given, I have an experiment I would like to try if I had another life to live. It would be practical on small places like mine but probably not on large farms. I would try pasturing hogs on the paddock(s) that I wanted either to seed or reseed to pasture or to plant to corn. In other words, I would let my unringed porcine plows cultivate the paddock while they ate, being watchful lest they root up soil onto the electric fence and short it out. Hogs rooting up a paddock would be a marvelous way to disrupt the cycle of sheep or cow internal parasites. If corn were the crop to follow, I would then broadcast the kernels in the fallow soil, maybe drag a harrow (or a cedar tree) over them, spray a corn herbicide to hold back the weeds until the corn got a good start, and see what would happen. Not having corn in rows would surely not matter. Many farmers who use herbicides do not cultivate for weeds anyway, so there is no need for having the corn in rows. I know broadcasting corn will work because innovative farmers in Minnesota did it in the fifties—thousands of acres broadcast by airplane. They found that they could harvest it okay with their combines even though the corn was not in rows. Planting corn in rows is just another of those things we keep on doing because we can't think of not doing it. (Actually, the corn industry is now talking about corn in rows eight inches apart, which is the equivalent of not having rows.) While the pigs plowed they could be providing themselves with their own free fuel and gaining weight, too.

So unimaginative is modern agriculture's view of itself that not one single study has ever been done to ascertain how well hogs might gain using as a supplemental feed the zillions of earthworms that a healthy soil generates. Don't laugh. It is a fact that an acre of good soil planted to legumes can contain two tons of earthworms in the top seven inches. That's a lot of protein supplement. On our farm, judging from the number of those terrible Japanese beetles emerging from the soil every summer, there's another two tons of beetle grubs that hogs could eat.

When the corn was ready, I'd pasture it. It would be sweet corn, so I could enjoy some of it, too. I know from experience that livestock like sweet corn fodder better than field corn fodder. I turned sheep on sweet corn once and they ate the stalks literally right down into the ground. The ears of sweet corn are smaller, so there would be less grain and less possibility of overeating. And because sweet corn never gets as hard as field corn even when dry, it should be more chewable and digestible. Milling it would not be necessary.

Okay, so I'm getting as wild as the missionary talking to the Illinois pioneer. But what if in a few years there is no cheap fuel anymore? We already have "chicken tractors" made popular by Andy Lee and Pat Foreman in their book of the same name. Why not pig plows?

But to consider pork production from a more "practical" point of view, it is true that hogs can't utilize forages as well as ruminant animals because they don't have multiple stomachs. Therefore, they are not thought of as pasture-grazing animals even though they love to graze pasture. As any hog producer knows, they gain weight more quickly on grain going through their straight exhaust systems, and the more grain the better. Grain and soybean meal can now get a hog up to 240 pounds in five months or sooner. But is that the most economical way to produce good pork? No, not when you consider the known *fact* that when you bed down hogs being raised on concrete and being fed a high corn/soybean diet, they will voraciously eat the bedding—

straw, cornstalks, whatever—until they have satisfied their dietary need for fibrous roughage.

There are some "crackpots" who are trying to serve the developing market for 60- to 80-pound roasting pigs for backyard barbecues. A look at the arithmetic tells an interesting story. An acre of corn at two hundred bushels per acre will fatten about eighteen pigs to 220–240 pounds each, the preferable market weight today. You have to add the cost of soybean meal and feed for the sows before the hogs were weaned. If my figures don't suit, use your own. The outcome is still going to be surprising to those who think corn-fed is the only way. An acre of alfalfa pasture will support two sows and their litters, according to Mr. Stevenson in the example above, but I know we were doing four sows and litters in Minnesota, so I will propose three sows and their thirty-six nursing pigs up to barbecue size. I would argue that the *net* income from those thirty or so succulent barbecue pigs could at least equal if not excel the eighteen pigs going to market as fat hogs, because the thirty could be sold at a much higher price per pound. (Actually, it would not take a much higher price because about half the time today farmers barely break even on market hogs.) And there is absolutely no reason the smaller hogs requiring only a fraction of the corn now fed today, if any at all, could not be processed and sold in the mass market as pork just as good as fat hog pork.

Entirely out of the question, the meat packers respond. It is much more economical to slaughter and process one 240-pound hog than three 80-pounders. That's correct (and proves that the debate is about economics, not diet or nutrition), but that argument ignores the fact that the meat industry regularly butchers millions and millions of 5- to 10-pound chickens and still manages to sell chicken cheaper than pork.

The pork market is not a free market. It is a market held captive by the corn industry. Free it and we might be amazed at what pasture farmers could come up with in the way of good, healthier pork at a lower cost to the consumer and to the environment.

12

Chickens, Ducks, Geese, and Turkeys Love to Graze

Those "chicken tractors" I mentioned in the last chapter are an excellent application of garden farming to grazing. When new blood and new brains get interested in horticulture and husbandry, new ideas come forth. It is important to note that this novel idea, like so many others, was adopted at first not because it was a profit-making venture but because it was a means toward someone's idea of greater utility in home food production. The typical chicken tractor operator doesn't give a hang about making money with it. He or she just wants to find easier ways to produce their own food. When the idea proved practical, *then* it eased into commerce. A chicken tractor, as I think every garden farmer now knows, is a movable pen in which chickens are kept while they "graze" last season's or next season's garden, lawn, pasture. The enclosed chickens in their floorless pen eat the bugs, weeds, and unharvested fruit and vegetables as the "tractor" is moved gradually from one spot to another. In the process, the chickens deposit their droppings as fertilizer and lay eggs. The concept is much like strip-grazing dairy cows on pasture, with the enclosure taking the place of the electric wire. On a commercial level, the chicken tractor becomes a mobile coop in which a flock of hens graze the grass under them or are allowed to go outside within predator-proof fences to graze. On both commercial and backyard levels, the "tractors" now come in all sorts of designs, and of course manufactured ones have begun to show up in garden catalogs.

After I thought I had learned all I needed to know about raising chickens, two little boys (my grandsons) and a little girl I have never met are teaching me more. Their fresh young minds see possibilities that I, brought up in traditional farming, was blind to. The grandsons made pets out of their chickens. They rendered those wire chicken catchers obsolete. When they want to catch one of their hens, they just reach down and pick her up. The hens like to attend our picnics. If we do not watch carefully, they will jump up and peck at the sandwiches we are eating.

When I was a child, chickens were always part of the scenery, but never did anyone think of them as lovable pets. Certainly none of us then would ever have written the kind of letters that a little girl named Chelsea writes to me from southern Ohio:

> Dear Mr. Logsdon,
>
> We now have 19 chickens! This is how it happened. After Buddy [the neighbor's dog] killed 15, we had 9 left. Then two were attacked by something and Buddy came back and killed 3 more. So then there were four (if Buddy hasn't come back since then). Then the man we get our organic feed from had chickens for sale. So we bought two laying hens, FH (Friendly Hen) and Little Red Hen. They are *really* friendly. FH will even come to me and let people she doesn't know hold her. Three others are Ellen and the other bigger one that doesn't have a name, and a baby. His name is Scruffy. FH, Little Red Hen, Ellen and the other one are all Bolvine Browns. Scruffy is part Bolvine Brown and part Aracauna. After we bought them we are back up to nine. On the very next day, the lady we buy eggs from called and asked if we wanted some hens. She gave us three Buff Orphingtons, three (old) Barred Plymouth Rocks, a Black Giant (she thinks so) a brown one, an Aracauna, and a young but very big and skinny rooster. None of them have names yet. I've been feeding them about three cups of feed a day (for 19 chickens). But the feed is 26 cents a

pound and they can eat bugs and dozens of tomatoes and go all over our land.

Sincerely,

Chelsea

And sometime later:

Dear Mr. Logsdon,

The number of chickens that I have keeps going up and down. Last time I wrote I had nineteen. Four roosters and 15 hens. But five hens weren't laying probably because they were so old. Then one of the Barred Plymouth Rocks died and we killed the other four that weren't laying and a Barred Plymouth Rock that was mean. Then we had 13. A little bit after that Uncle Joe gave us his hen because it lives in the garage and made a mess of everything. That gave us 14. But last night a Buff Orphington was missing. She still wasn't back this morning. She may have died because she was acting a little strange. She just wanted to stay in the coop. So now we have 13. They are: the two Hennriettas, FH, Little Red Hen, Charles, Henry, Homer, Scruffy, Bussell, Rusty, Cousin, Wilbur and Dotty. Bussell, Rusty, Cousin and Wilbur are Barred Plymouth Rocks. Dotty is a Pearl White Leghorn. She is small, too. We get 7–8 eggs from our nine hens.

Sincerely,

Chelsea

And later yet:

Dear Mr. Logsdon,

The Buff Orphington came back the day after she left but then she left and come back two more times. Then she died. She was old though. So now there are 13, (a baker's dozen). Charles chases the other roosters. Henry tries to get Charles' wives. Homer likes to eat and explore when there's no snow. Scruffy likes to eat and explore, also. He doesn't mind the snow quite as much as Homer. Henrietta likes to eat and stay in the coop when it's cold. Dotty is very skittish although she is getting calmer. She lays an egg every day even when it's only ten degrees F (her eggs are white). FH likes to travel. She'll go almost anywhere. Little Red Hen looks timid and she is sort of. She usually stays near the pasture at least. Bussell likes to go for walks with me. She limps when she walks. I think Buddy did something because she was the one that wasn't doing very well the last time Buddy came. Cousin isn't very friendly. Hilda is very pretty. She has bright black eyes and a very red comb and wattles. She is also friendly. They get all the table scraps except the meat. Cricket my dog gets that. They love apple peelings, old rolls and burned poppy-seed muffins. Most of all they love corn. Buddy came over one morning but the chickens were still locked in the coop. I haven't been letting them out until lunch time because it's been cold and it's much warmer when the door is shut. They don't like snow either. Even when I do open the door they don't go out except when I feed them and to drink. Their water is outside because if I put it inside they get straw in it. I give them about two cups of feed in each of two feeders. We are getting about 1–3 eggs a day.

Sincerely,

Chelsea

If you read Chelsea's letters closely, including between the lines, you know about all you need to know to raise a few hens. Chelsea is learning more about chickens, I daresay, than all the executives of Tyson Foods put together know. The reason I quote her at length is because she is giving convincing proof of a most important lesson of husbandry, particularly in pasture farming. Animals have individual personalities and you have to take that into account if you want to be a good husbandman. Science is made nervous by the idea that animals are individuals. A psychology professor once declared in a class I was attending that animals do not have individual psychological differences within a species. He insisted that only humans have personalities. When I would not quit arguing against that notion, he asked me to leave the room. I not only left the room; I left the whole stupid school.

I suppose as long as one stays within the abstract world of mathematics or even the molecular cause-effect structures of chemistry, scientific methodology is fairly straightforward and can lead logically to some valid conclusions. But in the real world of human and animal behavior, science can easily founder on the almost unlimited variables that come into play. Husbandry is an imperfect science. Agricultural scientists perform their experiments in test tubes or trial plots or in other environments where they can exclude as many variables as possible in an effort to isolate the one cause-effect phenomenon that they are studying. Generally, their conclusions hold up only until they clash with some other isolated cause-effect experiment. Why society in general puts so much faith in this kind of science is beyond me.

But, protests agribusiness, it is supremely foolish to suggest that we can feed the world with backyard chickens. (The arrogant old "feed the world" syndrome again. America can't feed the world; only the world can feed the world, and I wonder even about that.) My answer: We backyarders could do it as well as agribusiness can. There are 260 million people in America now. If just one-tenth of them raised an average of 100 chickens, that's about 2.5 billion chickens. If they raised an average of 200 chickens using Chelsea's methods, that's 5 billion chickens, enough for everyone in the United States, including Buddy. And 26 million young Chelseas would get an education in real life. And because the production of poultry would be dispersed, it would be much more dependable and safer than the concentrated production of a few huge chicken factories in case of war, economic collapse, or epidemic disease.

I can't resist an aside. The mindset that leads to consolidation in agriculture, so evident in the chicken business, has also taken place to an alarming degree in human culture, especially in consolidated schooling. Just as we herd more animals into confinement buildings, we herd more children into classrooms. Then we have little choice but to follow the rule of the chicken factory: one size fits all. And we justify both kinds of concentration camps with that all-American article of faith: It's cheaper per unit; we can't afford to do otherwise. Then we wonder why we must de-beak chickens and frisk schoolchildren for firearms.

In pasture farming, the fact that animals have different personalities reflects directly on how they graze and how they adapt to grazing. The grazier tries to select and breed for animals that have grazing personalities, so to speak, not lie-in-the-shade-and-bring-it-to-me personalities. So the husbandman must be artist as well as scientist. He must particularize management as well as follow general rules. As

Chelsea says, some chickens like to roam and some don't. Pasture farmers and scientists who have studied grazing animals in great depth in an effort to discover the rules and formulae by which grazing can be a profitable enterprise realize eventually that husbandry is an imperfect science. Andre Voisin, the assiduous student and scientist of grazing, was forced to admit (because he was a practicing farmer, not a laboratory scientist) that scientific guidelines only help so much. One of his final pronouncements in *Grass Productivity* he emphasized, so I will too: "THE PERSONAL CHARACTER OF THE COW UPSETS ALL OUR FIGURES" (91). Voisin understood that the infinite variability of natural behavior extends also to plants: "THE GRASS COMMANDS. IT IS NOT A CASE OF RIGIDLY OBEYING FIGURES; ONE MUST FOLLOW THE GRASS" (177).

All of which leads me to repeat the conviction that I am most criticized for in agribusiness circles. The biological world, not the economic world, rules the way we produce food. Farming cannot follow the economic principles by which the business of manufacturing is supposed to operate. Farming should not have to be performed completely within the profit-making strictures of manufacturing and money interest. Why this should raise so many eyebrows among those who think they are free enterprise farmers is beyond me. They all rely on government subsidies to survive today. The whole subsidy program, which in one form or another has been around since the arrival of the industrial revolution, is proof of the truth of what I am saying. Biology can't compete in the world of the assembly line. Forced to do so, it must have financial help. A chicken will grow like a chicken every time, not like exponential, variable money interest rates. If you put chickens into assembly lines, you are asking for trouble that not all the antibiotics in the world can avoid. What is currently called agricultural economics is a contradiction in terms. Economics as now taught is not about the biological values that govern farming but about trying to make cows and chickens and husbandmen act like accountants.

Therefore, I am hesitant about telling anyone that it is possible to make a good profit from farming, especially with chickens, in competition with factory farming. The only way I can see it work for the smaller producer is to develop a retail market and to process the meat or eggs yourself. And then be able to charge what pasture-raised, drug-free chickens are really worth—at least twice the value of factory chickens in my estimation. To make an entire living that way is possible, but it requires herculean effort. Small chicken producers in our neighborhood charge only five dollars for a broiler, butchered and cleaned. They should be charging at least ten dollars, if you ask me. Thoughtful customers sometimes pay them a bonus. I know one thoughtful lady who pays the five dollars and a homemade cream puff for each chicken. Now that's the spirit.

On the other hand, to raise pastured poultry for your own and your neighbors' tables or to have an operation that aims to make just a little money—say, a thousand dollars a year—is possible and practical. To produce just your own eggs is a snap. Any backyard that can provide a home for a dog will make a home for a couple of chickens. There is no farm so big or so small that it would not benefit from having a flock of hens. Hen flocks have disappeared from farmyards because they aren't "profitable" in an assembly-line world, a fact that should have no influence on the situation. If the flock supplies good fresh eggs and good fried chicken at low cost, who cares if it is profitable. Back before about 1945, when every farm had its flock, the enterprise wasn't very profitable either. One of the mysteries of life for me is the fact that thousands of families who insist that they are farmers won't keep at least a few chickens for eggs. The chicken coop is still standing, empty. Acres and acres surround the farmstead where the chickens could graze. Enough grain falls off trucks and out of harvesters in the barnyard to feed them most of the time. All the farmer would have to do is open and close the henhouse door every night and morning in order to have really good eggs.

Many houses in town used to keep chickens, too. (My parents kept a *cow* in town before they moved to the farm, and my son-in-law's grandparents kept a goat on their house lot in Cleveland.) When my Grandmother and Grandfather Rall were forced by old age to quit farming, they moved to town but took some of the farm with them, including a flock of hens. They did it the hard way, being unable to change their traditional farming habit. Instead of dividing their backyard into three or four parts for rotational grazing and keeping only six hens for all the eggs they needed, they kept a flock of about twenty-five in a coop with only a small run outside and sold the surplus eggs. (They needed the money about as badly as Bill Gates does.) That of course meant a buildup of manure that today's town society would not tolerate.

Chickens will eat just about everything except citrus. I imagine there are some that will eat even that—the northern oriole will eat oranges. I once butchered a chicken that had a marble in its craw. Being excellent scavengers, chickens have an advantage not often alluded to. They clean up grain and other feedstuffs around the barn and garden that would otherwise attract rats and mice. They are fairly good mousers, should a mouse be so foolish as to venture into their coop. A garden farmer acquaintance says that his chickens will even attack and kill pesky English sparrows that alight around their feed trough. I have watched my hens deftly peck flies off the faces of cows lolling in the shade, even pecking very close to the cows' eyes. The cows, being much smarter than humans understand, did not so much as move an eyelash.

Because chickens are good scavengers, they are next to perfect for pasture farming. They will eat the leaves off many plants and peck bugs off the others. They will eat many kinds of weed seeds; their tidbit brains somehow know that these seeds pack more nutrition than domestic grains. Livestock droppings are dessert to chickens. First they will eat some of the half-digested grain and grass in the manure. They will scratch an old, dried scat apart and eat the worms and bugs under it. Chickens, as well as ducks and geese, consume fly eggs and larvae in the droppings and also larvae of livestock internal parasites. Cagey graziers try to move fowl around with their other grazing animals for this reason. The grain monopoly says that chickens will eat only one-third of their diet as grazers and the other two-thirds need to be supplied by grain. That's just not so if the hens can roam like ours.

Oddly enough, I can find no firm conclusion from pastured poultry experts on precisely how much pasture can substitute for grain feeding in a commercial operation. The upper limits seem to be 30 percent, the lowest 10 percent. No conclusion has been reached because of the essential difference between a commercial pasture farm and a small-scale or backyard operation plus the difference between keeping chickens for eggs and keeping them for meat. But in any case, a commercial operation that can save 20 percent on its feed bill with pasture is saving a lot of money. To gain another 10 percent or more from pasture would require at least an acre per four hundred chickens and access to other land planted to grain for the chickens to eat on their own, an alternative that the commercial poultry business has not tried to my knowledge.

Chickens or any other fowl don't need corn in their diets. There is nothing magical about corn. It is just handy and, in our present culture, cheap. The Scots who pioneered pasture poultry a century ago do not have the soil or climate for much corn, but can they ever raise oats and barley. What grain they feed their chickens is therefore what is available, and the chickens do just fine. The main grain supplement for farm animals in Scotland is from barley after it has been malted and fermented into Scotch whiskey. That's what I call a wonderful economy.

Our dozen laying hens eat very little grain because they can graze over several acres of woods and pasture and find a great variety of wild plants and bugs and worms and weed seed grains, plus table scraps and meat scraps from butchering hogs and from carcasses of raccoons that I trap and kill. They won't eat grains at all when plenty of these other foods are available. On the other hand, the thirty Cornish Cross broilers we raise for meat, which are only six weeks old when butchered, are not good foragers, and they are terrible dodgers of hawks and coyotes, so we feed them mostly grain and let them "run" outside only in the last two weeks of their lives. They can't really run; wobble is more like it. I could force them to forage more, but I think the foxes would be the main beneficiary.

The serious drawback to grazing poultry is predation by wild animals. Or tame ones, as Chelsea has learned. Out on pasture, chickens are fairly easy plucking for Cooper's hawks, which are almost impossible to protect against. A man in our county who raises quail and pheasants has erected elaborate pens, about two acres worth of paddock entirely screened around and above to protect from hawks and owls and raccoons. Last winter during a big ice storm, the netting broke down and hundreds of pheasants escaped. But even without ice storms, such large netting enclosures might be practical for high-value fowl but not cheap chickens.

Coyotes, wolves, dogs, foxes, and other predators can end your chicken business unless the pasture is surrounded by electric poultry netting. If skunks, weasels, mink, and raccoons get into the coop, they can wipe out the flock. A mink did that to us once. Twenty hens slaughtered in one night after the mink found a place to dig under the wall. Predators seem to go berserk if they get into an enclosed building with poultry (like humans at a department store sale) and kill for the sheer delight of killing.

Commercial pasture production of poultry is going through a process of experimentation in predator control. I don't know if such control can ever be foolproof. But graziers are coming close. After the chicken tractor, Joel Salatin, Herman Beck-Chenowith, and others enlarged on the idea with bigger and more substantial but still movable pens. Progress brought not only better protection but systems to handle more poultry with less work. The first pasture pens would handle up to about fifty hens and were light enough to be moved manually. Because they had no floors, it was difficult to keep digging predators from getting into them. Anything added, like metal screening on the ground around the enclosure to discourage animals from digging, made moving the pens slower and more tedious. It didn't take much of a storm wind to blow those pens into the next county, either.

Next came sheds with floors. They are built on skids (sometimes on wheels) that require a tractor or team of horses to pull. The framing, usually of two-by-fours or two-by-sixes, is rather fragile but adequate. Roofs are usually of sheet metal. Some have sides that can be raised to afford shade or lowered for storm protection. These coops will not blow over as easily as their predecessors. There are as many variations in design as there are graziers. You can find literally reams of information about various designs on the Internet and in books and magazines.

Electric poultry netting is a fairly effective way to protect hens from all the Buddies of the animal world. I've heard of one grower who also stretches an electric wire across the entrance to his coop. The entrance is off the ground a foot or more. The chickens easily learn to jump in and out of the coop over the wire, but a dog or fox will put its front feet on the entrance preparatory to jumping in. At that point it comes in contact with the wire with its back feet still making an excellent ground for the electricity. End of problem. Doubtlessly many creative innovations with electric fence will be adopted in the future.

What is happening in building design seems to be a steady progression back toward the structures that poultry used to be raised in—that is, in permanent, well-built coops that can house several hundred chickens, but differing from previous factory-type facilities in that the poultry have access to pasture all around them, rotating from one paddock to another but going back into the coop at night. Chickens do not need to be trained to return to their coop if they have been penned there for a couple of weeks or more.

A permanent coop necessarily limits the number of poultry in each house to the practical availability of adjacent pasture. I personally believe this is the best solution. If you want to raise chickens or turkeys in sufficient numbers to make money at it, divide your pastures into units or what are sometimes referred to as cells, each with a permanent building surrounded by several acres of rotated pasture, each cell containing, say, two hundred birds. That way, the chickens or turkeys can go back in their predator-proof building at night, when predation is most common. This relieves the grazier of having to move coops around the pasture. As a rule of thumb, an acre of good pasture will generally carry one hundred turkeys or four hundred chickens. If you divide the pasture into paddocks, you can use all poultry, especially turkeys, to do the "cultivating"—that is, graze to nearly bare ground—before seeding grains and clovers by broadcasting. In concentrated numbers the turkeys or chickens will also deposit enough manure to provide all the fertilizer needed.

If I were to rate predator danger in terms of frequency of attack, my list would read: dogs, raccoons, coyotes, foxes, rats, feral cats (mostly in the case of chicks), skunks, weasels, hawks, and minks. Some animal lovers will consider me with loathing, I suppose, but raccoons need to be trapped and exterminated just like rats, because they are extremely destructive of other kinds of wildlife (especially bluebirds) as well as of poultry and crops. Human society's salvation from raccoons is distemper, which kills them in epidemic numbers when their population becomes dense. Ironically, farmers killing raccoons actually keeps them healthier.

Weasels are the most difficult predators to keep out of the henhouse because they can squeeze through any hole that a mouse can. Wherever there's any opening at all in our coop, I have covered it with hardward cloth.

I know people who really love wildlife who carry a rifle in their pickup for the specific purpose of shooting feral cats. Domestic cats gone wild take a severe toll on songbirds. They are the direct result of humans who are too lazy to take care of the pets they own, so they dump kittens in country ditches rather than taking them to the Humane Society for proper disposal.

The others predators in the above list do more good than harm, seems to me. Now chicken producers will be angry at me, too.

Our resident sharp-shinned hawks have tried often to grab our chickens and have only once succeeded. A hen can get very rambunctious when a hawk lands on her. One of our hens, Frantic, will never get caught even by a coyote. She is far too nervous and high-strung and suspicious of everything. At the merest hint of anything that might be construed as dangerous, she squawks loudly enough to warn all the chickens in the neighborhood and races for the coop. I saw a sharp-shinned hawk carrying away a squirrel recently, and that is good, because we have too many squirrels, too. We have had very few English sparrows on the farm since the sharp-shinned hawk took up residence. Hawks also kill rabbits, our worst garden pest, and lots of mice. Foxes prey on our squirrels and rabbits, too.

The coyotes will prey on young raccoons, groundhogs, and opossums. Minks rarely are a problem here. Two got into our coop because I had not built it tight enough around the foundation. The same was true of the skunk. Skunks are otherwise beneficial animals and ought not to be killed unless they insist on living in your garage. For roving dogs that neighbors won't keep on their own property, I have an air rifle with which I can sting them good but not harm them. They catch on fast.

Dogs and domestic cats are usually easy enough to train not to attack chickens. Our coop is divided into two sections so that we can raise broilers in one side and the old laying hens in the other. To train a dog or cat, I put all the chickens on one side and the pet on the other, with a wall of chicken wire between. They soon become friendly toward each other. In fact, our cat is so afraid of everything in the world that the hens chase him away from his own food.

A neighbor has a most novel way to avoid fox, coyote, and dog problems. His barnyard hens, of various mixed breeds, roost in the trees around his farm pond. When canines approach, the chickens fly up into the branches. This to me is just another example of how, if the animals have a chance, they adapt to the possible dangers of pasture farming. Another neighbor has chickens and ducks and geese wandering all over his barnyard, an amusing menagerie. They hatch out new broods in the barn more or less on their own. At night they fly up into the barn rafters, daunting even to raccoons.

So how do you breed chickens to do that? Bantams come by the ability naturally. Otherwise, I think you let nature do some natural selection. The chickens that don't learn how to fly up into trees and rafters are fox toast. I have made a mistake in this regard. For years, the young Rhode Island Red hens I was raising insisted on roosting in the trees outside the coop. I dutifully chased them down and put them inside. Now, if that happens I shall just let them alone.

To raise turkeys, which are good grazers, requires a sense of humor. My brother woke up one morning to find his turkey (headed for the Thanksgiving Day meal) on top of his house. Just sitting up there. Ho hum. Turkeys have a reputation for dying easily, like sheep, especially if kept on moist ground. If they do not feel like dying, however, you can leave them out in a rainstorm and they will gobble away in sheer delight.

Geese are such good grazers of grass but not of broadleaf plants that they have been used commercially to weed the grass out of strawberry and mint fields. An acquaintance of mine thought he would try that by putting half a dozen geese in a rather small strawberry patch. The geese freaked out at being penned in the garden and ran up and down the fence line until they had trampled the berry plants into near oblivion.

Geese can be very loud and so can guineas, even though both otherwise are practical fowl for pasture farming. It is not a good idea to try to raise either if you have near neighbors. The positive side of their noise is that guineas make great watchdogs, and both birds will sound the alarm far and wide at the approach of predators, sometimes actually scaring them away.

Raising pastured poultry commercially usually involves processing the meat yourself because the profit margins are so slim. If you go this route, understand that you are being paid for being a processor, not a farmer, and butchering is not what I call pleasant work. Also understand that you may have to get a license in some states to operate. Professional pasture farmers, like Joel Salatin, have campaigned somewhat successfully against such regressive laws, but you will need to check your local regulations.

We have good friends who process their own chickens and they make money, too. But killing and butchering thousands of chickens or hundreds of turkeys every year is a tough job. I butcher thirty broilers a year for our own consumption and that's enough. Slaughtering and butchering animals is to me the worst work in farming. I do it because I figure that if I'm going to eat it, I ought to process it. Also, I am leery of factory processing.

But do not let a dislike for butchering stop you from raising some backyard hens. You can always bury them in the garden when their laying days are over. To butcher a few is really not bad once you learn how. We always scalded chickens and pulled off the feathers. Amish friends convinced us to try skinning. Skinning is much easier. Just make a cut up the breast to the gullet and peel the skin off, feathers and all, almost like you were taking off coveralls. The wings generally require using a knife to cut away the skin from the wing feathers. I can clean and dress a broiler in fifteen minutes and so can you. But, anyway, there is in nearly every rural community someone who does custom butchering of poultry at a reasonable price. Give him or her a little extra money as a bonus.

13

Bluegrass, Ryegrass, and White Clover

Forages: The Science of Grassland Farming lists 360 "important" forage crops for the United States. Obviously, I can't address them all satisfactorily in one book. Instead I'll do readers a favor by ignoring most of them, which is what pasture farmers should do, too. I will concentrate on the ones that are widely used and point out the ones that, although not widely used, are important in particular regions or situations.

There is a terrible desire in pasture farming to find some miracle plant off in some obscure corner of the world that will grow through drought and snow drift and solve all our problems. Searching for miracle plants can be an absorbing undertaking, and I recommend it for those who can afford to spend the time or can con some grant-giver into financing the search. You could spend a fascinating lifetime that way and not yield success, because there ain't no miracle pasture plant.

Bluegrass or ryegrass when combined with white clover is almost a miracle, and I think the combination, all things considered, makes a most important and reliable pasture from the East Coast to the Plains, except in the Deep South. The only disagreement is over whether bluegrass with white clover is better or worse than ryegrass with white clover. Bob Evans, one of the leading lights of year-round grazing, tells me—and has on more than one occasion scrawled across pasture information he has sent me—that "bluegrass is a weed and has no place in Ohio." He knows I favor bluegrass. We are both very stubborn, so, likely as not, we will never agree. I suspect that on his steep, hilly southern Ohio pastures, bluegrass doesn't grow nearly as well as it does here on good corn ground in northern Ohio.

My first reason for being a bluegrass/white clover fan is economics. In most of the humid parts of our country, any piece of land where the soil has a fairly neutral pH value (6 to 7), is fairly well drained, and is in more or less full sun will result in a permanent stand of bluegrass and white clover if it is mowed repeatedly over several years. Sometimes the bluegrass/white clover will move in even faster. Seeding is not necessary. Many people won't believe this amazing fact or are in too big a hurry to wait the several years it might take for bluegrass/white clover to dominate the pasture. The seeds of these two plants are everywhere and remain viable for years, especially white clover. Both plants spread by stolons, too. Both plants have a tendency to take over wherever they find the soil to their liking, as you know if you have ever tried to keep them out of a strawberry or raspberry patch.

Don't take my word for the magical appearance of bluegrass and white clover when you mow competing weeds and tree seedlings. Well-known grazier-dairyman F. W. Owen, the same whom I've already quoted as believing that bluegrass and white clover make the best pasture, also says on his website (http://www.bright.net/~fwo/index.html): "Fortunately, any pasture in Ohio (or weed patch, corn stubble and old hay field) will quickly become dominated by bluegrass/white clover if grazed very short, and immediately protected from all grazing of re-growth . . . until it is six inches in height, and repeat the cycle forever." That sentence is really all you need to know about starting a good pasture in most of the United States or at least in the Midwest, Northeast, mid-South, and coastal Northwest. Some soils where this combo won't grow of its own accord need only a two-ton-per-acre application of lime every five years and/or better soil drainage.

Bluegrass is not even a native plant. It was introduced from Europe (one of the few good things introduced from Europe) and spread over the nation before the twentieth century. In less than a hundred years it spread from the East Coast through Ohio and Kentucky.

But most graziers prefer ryegrass over bluegrass to grow with white clover and invariably choose one of the improved white clovers like Alice instead of the naturally occurring white clover. Although not as permanent as bluegrass, ryegrass grows as early in the spring and as late in the fall. It is very palatable and digestible, easy to seed by broadcasting, and quick to establish itself. Improved ryegrasses and white clovers are regarded by graziers as higher producers of forage than bluegrass and naturally occurring white clover. Ryegrass's drawback is that even the perennial varieties do not last much more than four years, whereas bluegrass is forever.

In most ways, ryegrass is much like bluegrass. If it does produce more forage, something I am not convinced is true, the higher yields are offset by the fact that you have to buy the seed and run the risk of endophyte infection, which does not affect bluegrass. Humans are quite convinced that if something is free it can't be as good as something that costs money. Every advantage in nature has its downside. If ryegrass generally grows a little taller than bluegrass before going to seed and so appears to be producing more forage, it is also more prone to shade out the low-growing clover with it. So opting for ryegrass means spending more money on the new white clovers like Alice or Will, which also grow taller. A commercial farmer might profitably spend that money. A garden farmer might well stick with bluegrass and naturally occurring white clover.

Ryegrasses are generally divided into annual, perennial, so-called Italian varieties, and many hybrid crosses in between. Generally speaking, you can follow this rule: the easier a ryegrass variety is to get established, the less palatable it is. But they are all palatable enough. Annual ryegrasses are annual by definition, but because they will reseed themselves, they can be managed almost as if they were perennial. Italian ryegrasses last a little longer than annual ones but are hardly perennial, lasting at most three years. In traditional agriculture, Italian ryegrass has often been seeded by airplane on the bare soil of maturing cornfields for erosion control and pasture over winter. Where corn is grown in rotation with temporary pastures, this is a practice that pasture farmers should consider.

Ryegrass breeders claim that their ryegrasses build more organic matter in the soil than cereal rye. I would favor ryegrass anyway, because cereal rye can more easily get ahead of grazing or cultivation in the spring when you are busy with other chores. Most authorities and cows agree that ryegrass is more palatable than cereal rye, too.

The new interest in ryegrass as a major forage crop focuses more on perennial varieties and improved Italian varieties. This is especially true now that "endophyte-free" varieties are available. Endophytes are fungi that infect some grasses. The alkaloids in the fungi can cause, or are believed to cause, certain disorders in animals. Mention the endophyte problem among graziers and red alerts sweep the crowd. I am here to say, contrarily, that the endophyte problem is highly overplayed (as are most terror alerts). First of all, there is much debate about what endophyte infection in ryegrass and tall fescue really means. A good discussion of the problem appeared on page 1 of the August 1997 *Stockman Grass Farmer.* Just recently scientists discovered that there are "animal friendly" endophytes, too, which may be the best path to solving the problem.

Endophyte infection in ryegrass can cause "staggers," which seems to occur mostly when the animals have access to nothing else except tough old ryegrass during a drought. Staggers leads to loss of gain and other problems, but the toxicity itself rarely causes death and will clear up once good pasture is available. But animals with staggers, says the University of Missouri Farm Management Newsletter, are apt to drown accidentally, to suffocate due to crowding in gateways, or to succumb to other injuries associated with decreased mobility.

When I first bought ryegrass at the local farm supply store, all that was available was what was referred to as "lawn ryegrass" (this is the land of King Corn and Queen Soybean, and at that time pasture farming was viewed with suspicion by grain farmers, much as organic farming was when it first appeared on the scene). To my astonishment, the manager of the store said that lawn ryegrass (along with lawn fescues, which are now replacing bluegrass in lawns because they stay greener in August) were toxic to grazing animals. We both had to laugh—yet one more indication of the lunacy of human behavior. In order to get a lawn that most resembles a green carpet, you end up using poison grasses.

But something even funnier was involved. I had already seeded some of that supposedly toxic lawn ryegrass on my pastures and gotten good results. In fact, lawn ryegrass was the easiest plant to seed into a declining pasture that I had ever tried, better than the several perennial ryegrasses I have tried since then. I think maybe lawn ryegrass would germinate and grow on concrete. And what the sheep didn't eat went to seed and reseeded itself for awhile. Anyway, I was of course puzzled when I learned that it might be toxic because the sheep seemed fine while eating it. I read books on this knotty problem. Sure enough, common ryegrass is not necessarily endophyte free, although it was not clear exactly what that means. Fortunately, about that time I learned about F. W. Owen, the contrary dairyman I love to quote, who on his website argues that after bluegrass and white clover, lawn ryegrass, whether infected with endophyte or not, is the best all-around grazing forage. I quote him on the subject of lawn type ryegrasses: "Make sure [to use] a lawn type that doesn't shade out your [white] clover. It may be endophyte infected or maybe not, but it will survive northern Ohio winters. Agricultural type [endophyte-free] perennial ryegrasses are 3x as expensive . . . and won't live any longer." Then he goes on to say on another page that "the best pasture plants for dry matter intake per bite are bluegrass/white clover, then ladino clover, and third, lawn type perennial ryegrass." Every time there is a differ-ence of opinion about a grazing matter, my experience comes down on the side of Mr. Owen. I hope to meet him some day.

From my experience with ryegrass, I would not advise growing it for hay. It will make hay, but is hard to mow with a sickle bar mower and, like bluegrass, it goes to seed quicker than I can get around to cutting it. I've mowed it with a rotary mower, which shreds it so much that it is difficult to rake. However, chopped that fine it dries a little faster. A disk mower will handle ryegrasses better, but disk mowers are too expensive for my little operation. After it starts going to seed, which, like bluegrass, it will do quickly once hotter weather comes, the nutritional value of ryegrass declines drastically, although it is still good roughage to feed with lush legumes.

Because new developments in ryegrasses are constantly coming to market, you need to become acquainted with the pasture seed sup-pliers springing into business. I have found Byron Seed Supply, RR1, Box 92, Marshall, IN 47859 most helpful. You can learn a lot just by reading the company's product information guide.

Many graziers are coming to the conclusion that irrigating rye-grass/white clover pastures is profitable, even in humid areas where droughts are usually of short duration. I think so, too. It is also profitable to irrigate bluegrass/white clover.

Whether you go with common bluegrass and white clover or with improved ryegrasses and improved white clovers, it is the combination of the grass and the legume that works the magic. In both cases, the grass and legume have a symbiotic relationship to each other. The white clover, like all legumes, takes nitrogen from the air and transfers it to the soil via Rhizobia bacteria, to the tune of two hundred pounds or more of nitrogen per acre. The nitrogen attracts the grass like honey attracts bees. The grass grows vigorously until the nitrogen that has been quickly available is more or less depleted. While it is growing vigorously, the grass tends to outgrow the white clover somewhat. But after much of the nitrogen has been "consumed" by the grass, it grows less vigorously and the white clover comes back with a vengeance to supply another round of nitrogen to the soil. Back and forth the two plants wage war on each other and so the pasture benefits. Both plants are needed for good milk or meat production. Neither of them alone will equal the combination. Some agronomists believe that the way the two plants get along, especially bluegrass and white clover, is more than just the nitrogen relationship, but they don't yet know what it is.

At any rate, once a bluegrass/white clover pasture is well established, you need not, and should not, add any nitrogen fertilizer. All you will do is reduce the clover stand in favor of the bluegrass. Ryegrass devotees don't always agree and may add fertilizer. On a permanent bluegrass/white clover pasture, I don't think you should add any fertilizer, much less a nitrogen fertilizer. Given time, the manure adds enough fertilizer.

I think that the real basis for the resistance of graziers to bluegrass and common white clover is a matter of appearance. Bluegrass and native white clover don't grow quite as tall as other grasses and legumes, so it is easy to conclude that they are not producing volume equal to the taller forage plants. But I have often observed that in an established permanent pasture, bluegrass/white clover makes a denser stand than ryegrass and improved white clovers. A cow or sheep can get more in a mouthful than it can in a mouthful of ryegrass. Both bluegrass and ryegrass, however, grow more densely than tall grasses like orchard grass. Also, regrowth is faster, so you can often get two or more grazings from bluegrass or ryegrass/white clover, where you could get only one of the taller grasses and legumes.

A permanent bluegrass/white clover sod weaves a dense root system and a dense sod that will eventually blot out weeds better than ryegrass and improved white clovers. Sheep and even cattle and horses can walk on a bluegrass sod except in the muddiest seasons. F. W. Owen says that cows trashing bluegrass when the sod is soft don't really hurt anything. The clover will grow strongly in the hoofprints and in that way compete better with the grass. I think he's right. Once I drove a tractor on my pasture when it was slightly too soft. In the spring, new white clover clustered in the wheel tracks as if I had planted seed there. As for blotting out weeds, a bluegrass sod will, along with grazing and clipping, control even Canada thistle in a permanent—that is, undisturbed—pasture.

Another reason that the grazing fraternity doesn't give naturally occurring bluegrass and white clover much attention is that seed companies can't make any money from them. The two plants are there, free, from nature, so it doesn't much pay commercial companies to trumpet their advantages.

The weakness of bluegrass, ryegrass, and white clover is that they quit growing in droughty weather. For a month or sometimes longer, in August especially, you can't rely on them for the kind of pasture that increases meat or milk production unless rains are bountiful. In dry August, the grazier has to have deep-rooted legumes like alfalfa and red clover or cereal grains ready to graze. Or warm-weather grasses, which I will get to later.

Another criticism of bluegrass and white clover is that they don't grow tall enough for traditional haymaking. But since the idea of grazing is to limit haymaking as much as possible, what's the problem? For the garden farmer on tiny rotated lawn-pasture plots, there is no problem anyway. Just "make hay" with the lawn mower. You can let the clippings dry and rake them or bag them up for hay. My sister feeds lawn clippings to her horse. In plastic bags, the green clippings will store like balage, too. I use a rotary mower behind the tractor to mow bluegrass/white clover that is getting ahead of the grazing animals. Then I rake it up and make haystacks with it. Usually, these cuttings are past the optimum stage of palatability and nutrition, but the animals eat them just fine in winter or when they are eating lush legumes.

I am sure that bluegrass/white clover should be the choice for garden farmers who don't want to graze as intensely as commercial graziers and who need a permanent pasture that they can handle easily in their spare time. With an arrangement of very small paddocks with few animals, one or two paddocks can be spray irrigated as we do with lawns. Then you have super pasture.

Many grasses, including ryegrass and bluegrass, have a very interesting characteristic. Crossbreeding can change the grass, however slightly. In bluegrass, this characteristic makes it difficult to isolate cultivars to produce new varieties. But since bluegrass has been crossing in the wild for centuries and still reproduces itself fairly faithfully, we can assume it will continue to do so. Bluegrass also has a natural variability of chromosome numbers. It can reproduce viable embryos without having sex (to put it in terms we all understand). It can change morphologically as well as physiologically. What this means is that your bluegrass may not be exactly the same as my bluegrass. Fascinating. It explains (I think) why I have a wild bluegrass that spread across my lower paddock as soon as I gave it a chance to do so and that has an unusual ability to green up more quickly than bluegrass on other parts of the farm.

Two rules you need to follow with bluegrass and ryegrass. First, you should not let either go to seed and mature its first big spring growth. If the animals can't keep it down in early summer when it is growing fast, mow it and make hay out of it. Or mow it and leave it for adding organic matter to the pasture. The earthworms will love you for that. This will also set back the weeds that animals didn't eat. When or if the weeds grow back, they will be vegetatively more appealing to the animals. When the pasture grows back up again, more slowly as summer deepens, the animals can keep ahead of it, but if not, clip it again. Fall growth should not be mowed. Save it for winter pasturing. Don't let the brown color of dying fall growth fool you. It still has feed value. Second, after the pasture has been grazed down, and in this case you can graze down to an inch, it must be allowed to regrow to four to six inches tall before grazing again. You can tell your neighbors that you are using the Voisin method, and if you give that name a sort of French roll on your tongue, everyone will think you are a professional grazier, one of the saviors of the New World Disorder.

14

Alfalfa, Red Clover, and Ladino Clover

After Little Dutch white clover and its improved big leaf strains, alfalfa, red clover, and ladino clover might be called the Big Three of hay and pasture farming in most of the humid United States and even in some regions under irrigation. Each has its own characteristics, but I group them because they are grown, harvested, or grazed about the same way. They are managed more or less as perennials in the North, but in the Deep South, red clover and ladino are usually grown as winter annuals. Alfalfa is the acknowledged "queen of the forages" in the Midwest and West, but you might not think so if you have to contend with alfalfa weevils or leafhoppers. In our northern Ohio climate and soil, red clover is more queen to my way of thinking, and ladino at least a favored princess. In the past, these three legumes were primarily hay crops and seldom pasture crops. In pasture farming, they are still mainly hay crops but are more and more used for pasture, especially in drought times and winter. When our family was milking a hundred cows, we "grazed" alfalfa by chopping and blowing it into a forage wagon with a field chopper and tractor every day during the growing season and hauled it to the cows kept in dry lot. From a grazier's point of view, this practice, still common, is close to lunacy. Fuel is no longer cheap and electric fence for grazing is a whole lot cheaper than tractors, forage wagons, and field choppers.

These legumes, unlike common white clover, have to be introduced into your pasture. Introducing them means sowing them on old sod pastures or fields hitherto cultivated to annual crops. To the cultivator's mind, that means tearing up the soil surface with plow, disk, or field cultivator, working up a fine seedbed, and then planting the desired forage. This kind of mechanical seeding is rarely necessary, even on old sod pastures, and, of course, in many steep or rocky fields it is not even possible. When I first read what Bill Murphy, one of the leading interpreters and practitioners of the Voisin method of pasture farming, had to say about renovation, I was elated. He was presenting good evidence of what I had been observing, even though it ran contrary, culturally and agriculturally, to popular wisdom.

"Voisin was right!" writes Murphy in his *Greener Pastures on Your Side of the Fence.* "Pasture renovation won't make up for defective grazing management. In general, to improve your pasture, don't plow, cultivate, or kill the pasture sod. Test pasture soils and plant tissues and correct major soil fertility and pH problems. . . . Overseed by frost seeding or using a sod seeder to thicken swards and introduce new grass, legume and forb varieties."

On another page, Murphy says: "Conventional renovation practices essentially are attempts at transferring field-cropping techniques to a pasture situation. *They ignore the fact that a pasture is a different ecological environment compared to that of monoculture field crops grown on plowed and cultivated land.* Destroying a pasture sod and plowing and cultivating soils drastically changes conditions in soils and disrupts balanced relationships among the organisms that live there and make soils alive" (italics mine). I believe that if garden farmers and commercial graziers don't become convinced of what Murphy is saying, agriculture can never become a truly sustainable endeavor.

I prefer to call frost seeding "broadcast surface seeding," since frost is not technically necessary but just a good way to get good seed/soil contact and because winter seeding almost always guarantees plenty of moisture in the spring when the seeds germinate. *Sowing* is the right word since its original meaning was "scattering seed on top of the ground," but cultivation farming has appropriated the word to stand for planting seeds in the ground. Sowing is how you "plant" legumes and most pasture grasses.

The principle behind surface seeding can't be emphasized too much because it has all but been forgotten or ignored by the cultivator mindset in farming. The natural way that a seed sprouts most of the time, and the way legume and grass seeds sprout all the time, is by falling on the ground, coming into firm contact with the soil through rain or snow or freezing/thawing, sprouting when soil temperature warms up, putting a root down into the soil, and growing. Legume and grass seeds (as well as most weed seeds) are meant to sprout *on top of* the soil, not *in* the soil.

To convince yourself that planting seeds underground is not the rule of nature, study a white oak acorn. It falls from the tree in the fall and very soon sends a root sprout into the soil. Then it waits until spring to grow. All that poppycock about squirrels inadvertently planting acorns by burying them for winter food is a myth. Acorns not only don't need to be buried but grow better if they are not. The same is true of black walnuts and hickory nuts—in fact, all tree seeds. When they get solidly set on the soil surface over winter, they will sprout and grow like weeds.

As a second way to convince yourself that seed germination is more about the seed coming into firm contact with soil, not with being buried in the soil, examine closely how a no-till or a sod-planting drill works. There is no "fine seedbed" churned up before planting. The planter makes little grooves that penetrate just under the sod or soil surface residue and presses the seed *very* firmly against the side of the grooves. Although it may appear that the seeds are buried in the soil, close examination will show that the seeds pressed into the side of that little groove are actually almost exposed to sunlight, with only the most minimal amount of surface residue or soil duff actually covering them. Covering the seeds has one advantage. If moisture is minimal, the seeds have a better chance of sprouting *all at the same time and quickly*, which is important to grain farmers but not pasture farming. Legume seeds for pasturing or haying can sprout and grow over a goodly length of time without hurting the ultimate stand. In fact, some seedings may not show up strongly until months later. This ability can be an advantage, in fact.

With alfalfa, red clover, ladino, and other legumes, the books agree that the proper way to plant them is *on top of* bare soil or soil that is partially bare. But then the writers go on to instruct differently. This was especially true in the early days of pasture farming in the 1940s and 1950s, when they were still enamored with beating the soil to a pulp to prepare a seedbed and overfertilizing with chemical fertilizers. One of the great grazing books of that era is *The Pasture Book* by W. R. Thompson. It is aimed particularly at southern grass farmers but is full of details about grasses and clovers and cereal grains for grazing for anyone seeking practical knowledge about pasture farming. The author even understood 'way back then that the management of lawns and pastures is really the same thing. For all that, Mr. Thompson, true to conventional farming, believed firmly in beating the soil to death before planting anything. Since he knew, and so said, that legume seeds needed to be broadcast on top of the soil, not in it, his seedbed preparation instructions follow a curious, self-contradicting kind of logic.

He leads right off with a declaration that he presumes no one would dare to disagree with but which is directly opposed to Voisin and Murphy: "One of the most important pasture jobs is seedbed preparation . . . The right way to prepare a pasture is to do a good job of disking or plowing and disking . . . It will pay to disk sods thoroughly . . . in both directions, cultipack . . . and seed clover on top. Harrowing land helps prepare it for seeding before cultipacking." But he also says: "Never seed clovers on a loose seedbed. Cultipack or let rain settle the soil first."

Come again? First you loosen the soil and then you try to firm it up again before sowing seed. He repeats the advice in several ways, sometimes insisting that two runs of the cultipacker be made over the prepared seedbed before planting to firm up the soil sufficiently. In other passages he calls for seedbed preparation to begin two months before actual planting with several diskings so that there is plenty of time for rain to settle the soil firmly. He seems not to have asked himself a perfectly logical question. Why not just surface-seed without loosening the soil and let the rain do its germinating work? Cultivation makes no sense at all except to set weeds back a little, though in a pasture situation the animals will graze them anyway. I can understand Thompson's solicitude for the cultipacker in the South because of the importance of trying to preserve moisture for unnaturally late summer sowings there, but that seed is not going to germinate until it rains anyway, whether it is on top of the ground or half an inch in the ground. Loosening soil to make a "fine seedbed" is particularly risky at this time of year, because it can cut off capillary action drawing up water from below, if there is any there to draw.

Forages, cited earlier, contains a chart that indicates that legume seeds sown on top of the ground do as well as the same seeds planted shallowly in the soil. Only in a few instances did planting in the soil get better germination. But the experiment from which these statistics were drawn was from seedings in Ohio on May 18! That is two months after the right time to broadcast legume seed in northern Ohio. By May 18 we can get a siege of dry weather that would slow germination of surface-sown seed while the weeds forged ahead. In our farming tradition, March 19, the Feast of St. Joseph, is from time immemorial the date to plant clover seed. (You can actually sow it any time in winter or as early as November, for that matter. It won't sprout until warm spring weather comes and the longer it lies on the soil surface before that, the better contact it can make with the soil through rain, snow, or freezing and thawing.)

There is a slight risk involved (whether surface sowed or inserted into the ground), because if unusually warm weather activates the seeds to begin sprouting to what is called the "crook" stage, when the tiny sprout first emerges from the seed, the seeds can be killed if night temperatures fall below twenty-six degrees Fahrenheit. But, as often as that possibility was drummed into my head as I grew up, I have never seen it cause critical damage. Only a portion of the seeds reach crook stage at the same time and that stage lasts only briefly.

Forages says that shallow planting is better than surface planting, but then it does an about-face again: "On most soils in humid areas, practically all of the small seeded forage crops are best sown one-fourth to one-half inch deep, *or less*" (italics mine). And then the instruction goes on: "Few machines will place forage seed accurately at a one-fourth to one-half inch depth. . . . The most satisfactory machine for sowing [legume seeds] accurately on a prepared seedbed *broadcasts* [my emphasis] them between the sections of a corrugated roller with narrow corrugations. All drills, even the special 'grass-seed' drills, place forage seed too deep on loose seedbeds." It seems obvious to me, all things considered, that a $69 hand-cranked broadcaster is a better choice for sowing legume seeds than a $15,000 grass seeder.

In any event, if you can't rid yourself of the cultural addiction to tearing up soil into fine seedbeds, at least cultipack this loose seedbed *before* broadcasting the legume seed, not after. All experts agree with that. I made the mistake for several years of cultipacking afterward. I just could not believe that it was not a good idea to mash that seed down into the loose seedbed with a cultipacker or similar tool after broadcasting. In doing so, I was pushing many of the seeds deeper than half an inch into the soil and they weren't coming up.

To demonstrate my contention about surface seeding, I sowed clover last spring in last year's cornfield without disking or any soil preparation at all. I broadcast red clover and ladino seed in late April (too late) on bare ground so hard that the seeds bounced off it like tiny basketballs off a hardwood court. Rains did not fall immediately, so the weeds got a head start. I was sorely tempted to tear the whole mess up but decided that I might as well see what would happen. I mowed the weeds that threatened to shade out the tiny clover plants, and slowly, in May, with abundant rain showers, the clovers began to appear among the weeds. I mowed again and then pastured the plot for a few days. The sheep loved the weeds sticking up above the clover. Slowly the clovers took over. In August I cut the stand of clover and weeds (the field still looked awful) for hay. In a week, the clover, mostly the red clover, sprang back up, but the weeds did not. I had an almost weed-less stand for fall or winter grazing and for hay the next year.

All the instruction manuals advise you to plant legume seed in-oculated with Rhizobia bacteria to enhance the legume's ability to draw nitrogen from the air into the soil. This nitrogen fixation is the work of bacteria nodules on the legume roots, which you can see if you inspect closely. The nodules are pinkish in color. Since inoculated seed costs only a little more, and is easily available from many seed dealers, the grazier would be wise to take advantage of it. The effect on growth is sometimes dramatic, and of course results in more free nitrogen in the soil. Inoculant treatment is entirely organic.

Alfalfa

The dairyman whom I worked for in Minnesota years ago often said that he could "make a good living off any farm that would grow good alfalfa." He was speaking of dairy farming, but I think his observation would be true of most kinds of grain and livestock farming. Alfalfa loves a well-drained, neutral (pH 6–7) soil that is somewhat light and silty rather than heavy clay. It does particularly well on the dry but fer-tile soils of the West and on naturally well-drained soils of the humid

Midwest. If heavier clay soils are underlain with a good tile drainage system, and the soil pH level is brought up to 6.5 with lime, as must be done on most of our land in this part of Ohio, alfalfa will do well, but grudgingly. About the only way to find out on a particular field is to try it. Alfalfa came late to our neck of the woods, red clover being the preferred forage up until about 1950. I distinctly remember when the first alfalfa came to our community. An uncle planted it and it did wonderfully well. Everyone else had to try it and was surprised when it did not grow quite as rambunctiously as it did for him. The reason, soon understood, was that he had wisely chosen for that first planting a field of light, naturally well-drained, silty soil, not characteristic of most of the farms roundabout.

Alfalfa is called the queen of forages mostly because it will out-produce other legumes. It will yield five tons of hay or more from three cuttings, and, on the best soil, yields of ten tons per acre have been recorded from four cuttings. But it takes lots of extra fertilizer to get such high yields. More fertilizer equals more money spent. In nature, there is rarely any instance where one forage crop species really pro-duces "more" than another just because of its innate abilities. It de-pends on how much money you sink into the crop. If you drain, lime, and fertilize for a ten-ton yield of alfalfa per acre, you might get it. But if weather allows you to take off only five tons, or if the part of the ten tons gets rained on during haymaking, reducing its value considerably, you might lose money trying for high yields. Moreover, if you apply muriate of potash heavily, the usual form of potash fertilizer, in order to get super yields, organic farmers claim that the alfalfa will not be as palatable or digestible. Working with nature is a chancy occupation in which the clichés of business, like "it takes money to make money," are not always true. That is why there are so many instances of business hotshots getting into farming and losing their butts. Newcomers to farming, reading about alfalfa's greater potential for high yields, need to remember the downside when they see crusty old contrary farmers sometimes preferring the lesser-yielding red and ladino clovers.

Nevertheless, where alfalfa grows well, and the soil has good phosphorus and potassium content, it should be the first choice of the grazier for one reason above all: it will last four to seven years without a reseeding. This is a great advantage in temporary pastures where the grazier would like to maintain a rotation of five years or more of hay, pasture, and grain.

The first cutting takes place as the plants begin to bloom, in late May here in northern Ohio. When the stand regrows to bloom stage, another cutting can be harvested, about a month later. Then, where hay production is the priority, the stand is cut again when it grows back a third time. Highly fertilized stands might grow back for a fourth cutting, but in most cases three are preferable, with the regrowth from the third allowed to store nutrients in the roots for the next year.

In pasture farming, the first cutting is usually made into hay, but it could be grazed. My first cutting is always infested with alfalfa weevil, and I graze it when the weevil larvae are at their worst and the alfalfa is tinged yellow from weevil damage. The sheep eat the larvae along with the alfalfa, and the weevils are inconsequential in the regrowth. I do make hay from the second cutting. The third cutting in late August I almost always graze because that is the time of year when the bluegrass/white clover is at its lowest ebb. I make sure grazing alfalfa ends by September so that the plants have time before the first killing frost (around September 20) to make a healthy regrowth. This is an important detail. The plants need that September regrowth to store nutrients in the roots, so that in the next year the stand grows vigorously again. This technique also favors good grazing because the fall growth is then being "saved" or "stockpiled" for late fall and winter grazing. Generally speaking, in our climate alfalfa will turn brownish by late December, but unless it is covered with heavy snow or ice it will remain upright well into January for grazing. The fact that it is no longer green does not mean that it lacks appreciable nutritional value, as studies have shown. In fact, even frozen alfalfa (like most forages) retains fair nutrient value, as studies at Cornell have demonstrated. Cattle and sheep will graze it readily, as we learned last winter to our embarrassment when our sheep got into the neighbor's field.

There has been much debate over whether it is good practice to graze off this stockpiled alfalfa in winter. As I have discussed earlier, traditional belief holds that alfalfa needs that winter "cover" to insulate against winter freezing and thawing and subsequent frost heaving of the alfalfa roots. My experience, verified by other graziers, does not support this belief. By the time the most pronounced alternate freezing and thawing occurs in February, the "cover" of alfalfa plants has all but wilted away to nothing, leaving nearly bare soil exposed. The dead alfalfa residue offers no protection from frost heaving. You might as well graze it earlier, before the snow beats it into the ground.

Alfalfa's main advantage for the grazier is that it withstands drought better than any legume and so makes excellent emergency pasture. It will seem heresy to some, but I think that alfalfa as high summer pasture is better than any of the warm-season grasses being heralded for that time of year. Be sure to observe the normal routines to avoid bloat as described earlier.

There are many varieties of alfalfa on the market. Some are touted as being developed especially for grazing in that the crowns of the plants, from which new growth emerges, are lower on the stems and so not as affected by animals' biting off the stems. I have not seen much difference. I experiment and go with alfalfa varieties that do well in my soil, whether high crown or low crown.

Whether new varieties are that much better than old ones, I leave to your judgment. The ones that have some immunity to weevil and leafhopper attack are currently very popular, but I believe they are less palatable for animals, too. Throughout the seed business, there are rather vainglorious claims made for every new variety that comes along. Those who think that new varieties are going to make up for lack of good soil fertility practices and grazing management are in for disappointment. At any rate, it is useless for me to recommend varieties because even newer ones will be out before this book is. Experiment on your own.

Medium Red Clover

Because in more recent years agriculture has been enamored of the advantages of alfalfa, red clover hasn't received nearly as much attention. This is unfortunate because for much of the Corn Belt and the entire Northeast down into the mid-South states, red clover is a very desirable legume, growing better than alfalfa on tighter heavier soils and in colder, moister climates. But red clover rarely yields as much hay per acre, so in an agricultural economy ruled by quantity, alfalfa wins. I don't think that humankind will ever realize that an agriculture ruled by quantity rather than quality is not a good thing. It just isn't in our bones not to strive for higher yields even when higher yields do not mean higher income.

Last winter, on the last day of the year to be exact, we were experiencing a thaw. The temperature got up to fifty degrees. There had been a similar thaw earlier in December, but mostly the weather had been cold and snowy since mid-November. I was walking across the hilltop on my daily round of the pastures, and when I looked down at one of the lower paddocks where red clover was in occupancy, I stopped short. Was I imagining that there was a tinge of green to that soil surface down there? I hurried down the hill for a closer look. The red clover had started to grow. New little leaves were edging out of the crowns.

I could hardly believe my eyes, and I may not have believed them if a few days later I hadn't talked to acquaintances with connections at the Ohio State Research Station at Wooster. They casually mentioned that the researchers were all excited because they had observed that the red clover in their test plots was *growing in the warm spells of December*. So I knew I wasn't imagining things.

This discovery is not really momentous except to us grass-farming freaks. I had grazed that clover down to the ground in November. That it was growing after a few warm December days meant that it would grow more in warmer March days. And it did. When April turned cold again, and we ran out of hay because of the past summer's drought, I had four- to five-inch-tall red clover to graze. Although it is not good practice to graze red clover this early, I had little choice. It lasted until the bluegrass started growing again in mid-April. Forced to do what I didn't want to do, I learned something. The clover was not hurt by a few days of early grazing. It grew back okay for haymaking in June.

I think I can explain the ready response of the clover to the least little warmup in the weather. When I noticed how quickly red clover started to grow this spring, I got out a thermometer. Sure enough, when air temperature even two feet off the soil surface was, for example, sixty-five degrees Fahrenheit, the thermometer placed on the soil surface in full sun, with the bulb of the mercury just barely under the duff of old grass, read ninety degrees! When the air temperature was forty-five degrees, the thermometer, similarly placed, read seventy degrees. Great mother earth, with her body temperature of fifty-five degrees, had opened her bosom to absorb the sun's rays, soaking up the heat enough to warm the near-surface roots of the plants. The fact that I had grazed the plants down to the ground in the previous fall allowed the sun to warm the somewhat bare soil faster than if the surface had been shaded by thicker old plant growth The most remarkable characteristic of this early legume growth is that it is not killed by a spring freeze. It just sits there with a stiff green upper lip until warmer weather comes again.

Red clover is not bothered by any bug as serious as the alfalfa weevil or potato leafhopper in alfalfa. That alone ensures its continued residence on my farm. It does get a mildew disease, which is why it generally lasts only three years. The mildew doesn't really harm growth otherwise, in most years, and plant breeders claim to have new varieties that are more resistant.

Red clover as well as alfalfa will catch and grow even when broadcast on an old declining sod, especially if the sod is lacerated by sheep hooves before broadcasting. On heavier sod, I've used a disk in very early spring to make grooves in the sod for the seed. Set the disk straight so that it barely scratches in the grooves. Then broadcast seed. The seeds that fall in the grooves get good soil contact and sprout readily. This method is a whole lot cheaper than using a no-till drill.

But it is amazing how red clover will seed itself even down through grassy sod. One year I walked across our lawn with the little broadcast seeder slung over my shoulder. Unknown to me, the seeder had sprung a leak and the red clover seed was dribbling out. I soon realized what was happening and took care of the problem. Later in the summer, I noticed that there was a nice line of red clover plants across the lawn. It took me a little while to figure out why. The seed that I had spilled that day had taken root.

Eastern and midwestern farmers have always favored red clover because it is the legume that we can harvest for seed in our humid climate. (Most legumes harvested for seed are grown in the drier West.) In former years and often still, farmers have grown red clover for a crop of hay and then harvested seed from the regrowth. Graziers can take advantage of red clover's (and ladino's) ability to make seed in a humid climate. Some are allowing their red clover to head out and go to seed before they graze it. Then enough of the seed heads are knocked or trampled to the ground by the livestock to make another crop of clover. The surest way to get a new crop is to time second or third regrowth blooming until late enough in the fall so that seeds falling to the ground won't sprout until the next spring. Red clover sprouting in September or October might not overwinter. But if the second or, more likely, third cutting or grazing is stockpiled until winter, both the seeds that the livestock trample and the ones the animals eat and scatter on the field in their manure will sprout in the spring.

One of the surest ways to get a nice sprinkling of red clover for midsummer grazing in my permanent bluegrass/white clover pastures is to allow sheep to graze the stockpiled red clover and then run on the bluegrass sod. Their manure will be full of red clover seeds that, having passed through the animals, sprout readily and vigorously.

This practice is one way to deal with red clover's inability to last as long as alfalfa in a stand. Red clover is usually a three years and out crop: the seedling year, when it is good for one cutting or grazing in the fall; the second year, when it is good for two cuttings or grazings and possibly a third regrowth for winter grazing; and the third year, when it still makes a fair June growth but then starts to decline. But if the clover is allowed to reseed itself as described above, it can obviously last longer than three years.

The fact that red clover declines fast in its third year can be turned into another advantage. Usually in the fall of the third year, there is a considerable amount of bare or nearly bare soil that in the next year will be covered with weeds. You can plant winter wheat into this declining stand in the fall, or you can plant oats the next spring and get a fair stand without any cultivation. The clover seed from that declining stand will also sprout and grow another crop, as explained above, so you can renew the pasture entirely without cultivation.

Red clover is a good weed smotherer because it can grow three feet tall and be very dense. Some weeds do fight their way above the clover, but they lose out after the first cutting of hay is made. The clover grows back so fast that most weeds are overwhelmed. Then after a second cutting is made, the clover regrowth is often magnificently weed-free. This is rather true of a strong stand of alfalfa, too. Why buckwheat has gotten a reputation as a weed smotherer has always mystified me. Red clover and alfalfa are much better. The seeding rate of red clover or alfalfa is about ten pounds per acre.

New varieties of red clover are constantly being introduced because of interest from pasture farmers. The new ones are of course always heralded as better. Some farmers who have saved seed for generations believe their own homegrown variety has become acclimated to their particular soil and climate and is therefore better than the introduced varieties. I try new ones as a matter of course, but I can't see much difference. Weather and soil fertility mean much more than variety. Comments Jim Gerrish: "In controlled research trials here in Missouri, we found locally produced red clover 'ecotypes' to be equal to 'improved' red clovers, but sowing 'common' red clover from a distant region such as Oregon or Minnesota to perform poorly."

After being grazed down or cut for hay, red clover must be allowed to regrow to bloom stage before pasturing again. Like alfalfa and ladino, the regrowth period is roughly a month. As I have said, none of the three should be grazed or cut for hay in September unless you plan to renew the stand with something else the next year. They need the September growing period to put nutrients in the roots for the next year's growth.

There is also another species of red clover, mammoth red. I don't see it as having a place in pasture farming.

Ladino Clover

Ladino clover is big brother to the white clovers. It grows taller but is not as persistent. It won't last in a permanent pasture like Little Dutch will. Ladino, like all white clovers, is very palatable. It is grown more often for hay than for pasture. The stems of ladino are finer than red clover's or alfalfa's, so although it yields less tonnage as hay, more of the hay is edible than first cuttings of red clover and alfalfa, which often have thick, hard stems.

Ladino clovers will endure somewhat tighter, wetter soil than red clover or especially alfalfa. By the same token, it will not endure drought nearly as well as the other two, which is why I favor alfalfa and red clover. I like to sow ladino with red clover in lowland fields. Between the two clovers, I get a better stand all over the field. Ladino, like all the white clovers, has very small seeds, so you need only one to two pounds per acre as a seeding rate. I like to seed a mixture of two pounds of ladino and eight pounds of red clover.

Ladino may not sprout and grow as quickly as red clover or alfalfa, especially if broadcast after cold weather passes. Some of the seeds will not germinate until they have been through freezing temperatures. That is why sometimes you get a better stand the second year than the first. Or, interseeded into a declining red clover or alfalfa stand, you might not see a good take until a year after sowing. Ladino will winterkill in the far North, but often the injured plants will recover and the stand will thicken from volunteer seedlings. I have a hunch that common ladino clover is going to be superseded by the new big-leaf white clovers like Alice, a New Zealand clover, and Will, an improved ladino.

15

Legumes of Regional Importance

For my own farm and those of the northern Midwest and the Northeast, my current thinking is that the legumes discussed so far are sufficient. I stand ready to try others, but I have over the years tired of grandiloquent claims made for old and strange legumes. Circumstances may change my opinion of those legumes, but there is a lot of marketing wind blowing them along. I have tried some of them, but none so far seem any better than the ones already named. Nonetheless, there are worthy legumes that fit other regions or specific soil situations. Here are the most popular.

Lespedeza

Lespedeza is a southern to midsouthern legume used for summer pastures. The northern line of the legume's range runs generally through mid-Ohio, Indiana, and Illinois. Graziers such as Richard Gilbert in southern Ohio find it very useful, while a hundred miles north, here in northern Ohio, the weather is a bit too cold for it. Mr. Gilbert says Marion lespedeza on a neglected old cornfield grew marvelously for him. Then he sowed Legend lespedeza on a sour old fescue pasture and it, too, made a good stand. He waits now to see how well it reseeds itself.

Korean lespedeza and common lespedeza are annuals and are preferred in the northern half of the plant's range; sericea lespedeza, a perennial, is grown more in the southern half of the range. Sericea is a noxious weed in many states and becomes unpalatable quickly unless managed very aggressively. It should be avoided in almost all cases. There are other distinctions and divisions, but none of great consequence for graziers. If you live in lespedeza country, note what varieties other graziers are growing and do likewise. Whether or not lespedeza is annual or perennial is not of great moment either, because one of annual lespedeza's advantages is that it readily reseeds itself. In fact, it can be rotated with cereal grains and will continue to make a strong volunteer stand without any planting help from the grazier. The field is oversown to cereal grain in the fall or in the spring after a light disking. The lespedeza takes care of itself. In disking or other cultivating, care must be taken to keep the lespedeza seed near or on the top of the soil. From my experiences with broadcasting wheat, a lespedeza/cereal grain rotation could be carried on indefinitely without cultivation at all with good grazing management. Lespedeza comes on strong after the spring and early summer grasses begin to tail off, so it is prized as a succulent high summer pasture. Livestock do not bloat on it, and weight gain in all trials has been competitive with feeding grain on dry lot.

Although it is a rule in farming that "poor land makes poor clover," lespedeza will grow better than most legumes on poorer, light soils. On rich soils it doesn't compete with weeds very well. In fact, in general, the grazier thinks of lespedeza as pasture, primarily because grazing animals keep the weeds down and only secondarily because it can be used for hay. Although it can make good hay, leaves shatter easily out of the hay if it gets the least bit too dry before raking into windrows. The seed is a good source of protein supplement, so if grazing gets delayed until seed maturity, at which time the leaves are drying up, not all is lost. The livestock eat the seeds. Also, by then enough seed falls to the ground for the next volunteer stand.

Crimson Clover

Another southern legume, crimson clover is important as a winter annual in the southeast quarter of the United States up to about southern Indiana. It is used by southern graziers for winter pasture. Since, like lespedeza, it produces ample seed that volunteers for a new stand, it can be managed for continuous growth like a perennial, although it is most often used in rotation with other crops like cotton and grain. Continuous growth might encourage diseases to which crimson clover is subject. One of the legume's characteristics is that the seed will not sprout all at the same time after sowing, but over a period of time. So if the first seedlings to sprout die from dry weather, a not uncommon occurrence in late summer sowings, or if new seedlings are killed by late frost, not uncommon either, more plants come along from successive germination. Crimson clover also adapts to a wide range of soil types. I suspect that both these characteristics explain why a crop at home in the South finds favor, or at least used to find favor, in the northern tip of Maine where potatoes are king. Strangely enough, the legume will not survive farther south in Maine. Some is grown along the Pacific Coast, too.

Graziers sow crimson clover for winter grazing in late summer or early fall and then pray for rain. In the South, there is more insistence on firming the seedbed with a cultipacker because seeding for winter pastures during that season is more common than in the North. Since weather is often dry then, grass farmers will take great pains to get the seed to sprout. They may cultipack before and after sowing clover seed. As I have said, rain makes the seeds sprout and grow, not cultipackers, but if it makes a farmer feel better to cultipack, well, it won't hurt much, and feeling better is something every hard-pressed farmer appreciates.

Although plowing under legumes for green manure is foreign to the philosophy of pasture farming, and, in fact, is something a true grazier considers another example of agricultural lunacy, legumes are often planted on conventional farms for that reason. Crimson clover is especially used this way. The soil lover must be thankful. If one must plow and till the soil, which one must do for cash crops like cotton, better to use green manure than just chemical fertilizers.

Birdsfoot Trefoil

Birdsfoot trefoil is grown in the northeast quarter of the United States. It is to cold climates what lespedeza is to southern climes. It received a lot of attention in the fifties, but the popularity of this very fine-leaved legume goes in and out of grazing fashion. I gave it a good try, and what I have to show for it are a few scattered plants that persist throughout the permanent pasture. Its persistence is in fact its best characteristic. Other advantages are that it can stand a somewhat acid soil and poorly drained soil and it withstands drought very well. It is also very winter hardy.

Despite these good marks, I fail to understand why birdsfoot gets so much favorable press. Sometimes I think that ambitious plant propagators push new or unusual forages because they need something not so well known to justify new research and plant breeding. They are looking for new frontiers to conquer. I don't mean to belittle that ambition, because who knows when something new will turn out to be just what the future needed. But the truth is that beyond persisting in pastures if allowed to regrow properly, birdsfoot doesn't compare with the big three legumes and definitely not with white clover. It won't compete vigorously with other pasture plants or weeds. It doesn't make big yields of hay. When I made hay of it, the fine leaves mostly dropped off. Nor, despite all the press to the contrary, did my livestock find it especially palatable. The animals grazed it, but not eagerly.

Now all the northern champions of birdsfoot trefoil are going to be upset with me. So I will add that birdsfoot is better for frigid Minnesota than for somewhat milder Ohio.

Sweet Clover, Alsike Clover, Hop Clover, Black Medic, and Others

Minor legumes persist in farming, or at least in books about farming, because they have some specific advantages in specific places. Mostly their advantage is an ability to grow fairly well, or at least survive, on problem soils. I think it is a disservice to grass farmers to encourage these legumes. If your soil is such that it will grow only black medic or hop clover or alsike, then I doubt very much that you will make a success of grazing until you improve the soil so it can grow the better-known legumes. Experts who write agricultural books of instruction love to list and describe every blooming plant that has any possible connection with the subject they are discussing. Or they will make long, boring lists of numbers about oh, you name it—rate of gain of various animals on various pastures; nutritional levels of this or that forage in this or that situation; production figures over ten years of this or that. Yawn. These lists are next to useless. The authors aren't really trying to show how smart they are. They just know that editors are impressed all to hell and back with what looks like really knowledgeable writing. And the less the editors know about the subject, the more impressed they are. Also, beginners, trying to learn something about farming by reading, are impressed with long lists of names, descriptions, numbers, profits, and losses. For example, I am looking at a "table" in a book about forages that gives the "sweetclover acre yield of roots and tops in late fall of seeding years; average of 27 tests at 12 locations in Ohio." Does anyone really believe that a statistic showing the average number of pounds of sweet clover (3,580) that grew in one year in twenty-seven tests in twelve locations is telling the reader anything useful about sweet clover? This is a poor use of taxpayer money.

All the grass farmer needs to know about sweet clover is to avoid it. Sweet clover is great for improving soils that have been mismanaged. Its roots break up compaction in soils that have been pounded to death by poor farming. As grazing, it is not very palatable, and if you make hay from it, be sure that it does not get moldy. Moldy sweet clover contains coumarin, a blood thinner that can kill your cows.

For improving soils and for ground covers, the vetches are much better, in my opinion. The vetches, mostly grown in the southeastern quarter of the country (although hairy vetch will withstand below-zero weather), make decent hay and pasture, too, but are not much used that way.

My grandfather loved alsike clover. It will grow on poorly drained soils. He farmed land that did not belong to him. The landowner did not want to spend money on thorough tile drainage, which would have paid for itself in a few years. So grandfather had to grow alsike to scrape by financially. If alsike didn't exist, then the landowner would have had to put in tile. And then grandfather could have grown red clover and made a profit doing so. It is good, however, to sow a little alsike with your red clover. Then you can tell where you need better soil drainage.

Burclovers grow along the Pacific Coast and in the South. According to the books on forages, "all burclovers lack winter hardiness and cannot endure summer heat." Why in the name of all that is holy would anyone want to grow them?

Hop clover persists in pastures just about everywhere. Black medic persists in some. If these unusual clovers grow in yours, good. If not, don't worry about it. They have nutritional and possibly medicinal value, as the name black medic suggests. (All legumes have medicinal value; I suppose a case could be made for saying that all plants have some medicinal value.) It is fun to be able to identify these and other minor legumes in your permanent pastures, but, at least so far, trying to make important pasture plants out of them is not practical.

There are many other minor legumes. Beware of introducing something new no matter how loudly touted it is as a good pasture plant. Remember kudzu. It was introduced in the 1930s as the savior of southern agriculture. Oh, it is a wonderful grazing plant, no doubt about it. It has also been known to grow so rampantly as to totally engulf and pull down small buildings (I am not exaggerating). Kudzu became the scourge of the South. Opportunistic shepherds learned to profit from it by renting out their sheep to terrorized landowners to eat the stuff into subjugation.

Kudzu demonstrates that we do not live in a pastoral society. If we did, a plant such as kudzu might hold a place of honor until we ran out of enough sheep to keep it from destroying the earth. I know a farmer who is convinced that Japan introduced kudzu into America back in the thirties because the Japanese already had plans in the works to conquer us. Kudzu, says the farmer, is an excellent way to soften up the enemy.

16

Other Noteworthy Pasture Grasses

After bluegrass and ryegrass, other important grazing grasses are timothy, orchard grass, tall fescue, and bromegrass. Also important, but more or less specific to the South, are bermudagrass and dallisgrass. Wheatgrasses are widely grown in the northwest quadrant of the United States. In the dry plains areas, where bluegrass is all but unknown, native grasses of the bluestem family are mainstays. The grama grasses are adapted to warmer plains areas, but sideoats grama will grow throughout the Corn Belt and into the Northeast, too, if you can get it established. A renewed interest in native grasses in the eastern two-thirds of the country has led to experimentation there with the bluestems and grama grasses more often associated with the Plains.

Of the many other grasses that make fair grazing, I must, against my better judgment, mention foxtail, crabgrass, quackgrass, Johnsongrass, reed canarygrass, sudangrass, and sorghum/Sudan crosses. The first four are the terrorists of cash grain farming, and the others are of only limited adaptation to pasture farming. If I don't also mention redtop *(Agrostis alba)*, the bookish farmers will question my credibility as a grass farmer. Everyone who writes on grasses mentions redtop because it is discussed in earlier books, and so they think they must also discuss it. But hardly any farmer knows what it is and doesn't need to know. It was once a popular grass, but it's one I think you should ignore.

There will always be "wild" grasses, like yellow or barnyard millet, volunteering in your pastures. I keep an eye out for them in the same way I do for wildflowers. The animals graze them along with the other grasses. They seem to come and go without much ado. I have transplanted some of them in the lawn and grazed them with the lawn mower. It is interesting to watch how or whether they spread. Sometimes undesirable wiregrass and more palatable nutgrass gain a foothold in the pastures, but the animals eat them enough so they don't become a problem.

Sow grass seeds just like legume seeds. Most grasses will germinate and grow if surface-broadcast in the fall, winter, or early spring. Unlike sowing cereal seeds for cash grain, it is not necessary to get a nice, uniform, weed-free stand growing within a matter of days. Grass seed sown in the fall for pasture might not germinate until spring. If it is sown with a cereal grain or a legume, you might not see a good stand establish until the following year.

More than likely, over time, you will find that any of the grasses mentioned above that you have tried, and often some you never actually sowed, will appear in your permanent pastures. Although I aim to keep bluegrass, ryegrass, and tall fescue as the dominant grasses in my permanent pastures, timothy, orchardgrass, quackgrass, foxtail, barnyard millet, and several others are always around. That's good. But with all the grasses present, not to mention several legumes and several kinds of weeds, it is going to be very difficult to estimate the amount of forage that is growing in a paddock. It would be very handy if commercial graziers had a sort of slide rule or "grazing wedge," as they are called, that they could use to tell how much grass is available at any given time to help them decide how many animals could graze there for how many days before moving to a new paddock. As yet, none of the grazing wedge measuring sticks that I have read about differentiates one grass from another or mixtures of grasses and clovers. I think grazing wedges are just a game some graziers enjoy playing.

My argument against the slide-rule approach goes like this: There is so much dry matter out there in a paddock. Let us call one's grazing-wedge estimate X. The real amount might be X but most likely will be less or more than X—let's say either DM $= X - y$ or DM

$= X + z$. Whatever it is, you put the animals in and they graze. When they eat the forage down, you put them in another paddock come hell or high water or grazing-wedge guesstimates. Let's say your grazing wedge gave you an estimate of X number of tons of DM, which you decide to graze with V number of animals for W days. Do you move the animals based on the numbers suggested by the grazing wedge if the V number of animals ate all the DM in $W - 2$ days instead of your grazing-wedge numbers? Do you let them stand there and go hungry for the last two days because that's what your slide-rule predictions called for? Of course not. You use your eye and your experience.

Some manuals are full of numbers about DM and DMI and indexes of palatability, and TDNs and MEs and percentages of percentages, and corollaries of corollaries, and conversion rates and other strings of alphabet soup poured into vacuous algebraic formulae. The grazier reads awhile, his mind fogs over, and he runs to the pasture screaming for relief. You might as well try to inspire students to write a poem by teaching them calculus. And if you do wade through all those numbers and letters, and I have, the conclusion, as Voisin says (and he is the most notorious of all in joining jargon to formula and creating great clouds of fog: "**Figures are only guides; in the end it is the eye of the grazier that decides**" (his boldface, not mine).

Still, I would bow humbly to higher intelligence and try to use the slide-rule approach to grazing management if I could just find graziers who actually do it continually. They don't. When I ask them if they have been using their grazing wedges regularly, they give me that same queer look that I would get if I asked them if they have been regularly avoiding fattening food. They don't want to admit that they haven't, but doubt it's worth lying about.

It seems to me that the slide-rule approach to estimating dry matter, or timing of paddock rotation, or deciding the number of animal units to stock per acre, can't work very well. The first reason is that the measuring stick is usually based on forage height, and forage height is not nearly as important as forage density. A good thick ryegrass/bluegrass/white clover sward may contain more forage in four inches than orchardgrass contains in six. Or one stand of orchardgrass at six inches could contain more than another stand at eight inches.

Furthermore, height and density aren't the only variables. An acre of orchardgrass is not the same as an acre of ryegrass and white clover. And that is only the beginning of the complexity. There's not enough uniformity in any part of pasture farming for a slide rule to handle. There's no uniformity in rate of growth in varying weather; in the taste of pasture plants; in the animals' capacities to graze; in the amount of grass an animal can get in one bite; in the number of bites different animals will take before they lie down to rest; in the mineral content of different plants at different times in different soils; in the kind of weather during the grazing day; in the mood or intelligence of the person eyeing up the grazing stick. If there were, any damn fool could be a successful grazier, and I don't think any damn fool can be. The kind of rational mind that invents grazing wedges thinks that human and animal behavior can be reduced to mathematics. Phooey. Pasture farming is an art first and a science only second. No, that's not quite right. Pasture farming is luck first, then an art, and then a science.

Timothy

Timothy is by all accounts that I can find, including my own experience, a favorite grass of grazing animals. In my pastures, sheep and cows will always eat it readily. Timothy grows best in heavier clay soils in moderately cool, humid regions, notably in the upper Midwest and Northeast and Canada. It grows well where red clover grows well, and the two are most often grown together for hay. With the popularity of other grasses, like ryegrass and orchardgrass, the historical death of timothy is continually being announced but never actually happens. It was, and is, considered excellent hay for horses because it is higher in energy value than most grasses. As the use of draft horses declined in farming, so did the acreage planted to timothy. But, like bib overalls, timothy continues to endure and even increase in use with the advent of European varieties that appear to mature later in summer. Good timothy hay right now sells just as high to horse farms as good alfalfa hay. Just as amazing, and amusing, is that with the penchant for timothy increasing again with horse owners, I have seen poor timothy hay sell about as well as good alfalfa.

Timothy grows to about three feet in height and resembles cereal wheat both in leaf and placement of the seed head on top of the stalk. It is better for hay than for pasture, since it takes longer to recover from cutting or grazing than bluegrass or ryegrass, and so is harder to fit into a pasture rotation in combination with legumes. When it is the proper time to turn grazing animals back into bluegrass or ryegrass or legumes, the timothy, which has been seeded with these grasses, is usually not quite ready. So the grazing animals eat it off too quickly and it tends to die out. Nor does it spread by root rhizomes, so it does not form a solid sod. It does, however, spread by tillering and will continue to renew itself if managed properly—that is, grazed or cut before heading out—and allowed sufficient regrowth of six to eight inches before pasturing again. Because it is grown almost always with clover, the stand is usually managed for the clover, and the timothy suffers accordingly. The hay is cut after the timothy seed heads have begun to mature, which means the timothy is no longer at its peak of nutritional value or palatability and will not grow back as readily as when it is cut before seed heads develop. I once heard an agronomist say that when cutting a timothy/clover hay, the time of cutting should be gauged by the growth of the timothy, not the clover. Cut before the timothy goes to seed and don't worry about the clover, he advised. I think that is good advice, but often weather does not allow making hay early enough to catch the timothy at the right stage.

I have seen solid stands of timothy in the Northeast that produced remarkable yields of forage, as hay, silage, or pasture (or to harvest for seed), because it was managed in accordance with what was best for timothy, not clover. When seeded with red clover, which declines in its third year, the timothy often makes a strong growth in that third year and will continue to do so for several more years. This is not a well-known fact because traditionally a field of red clover and timothy is plowed at the end of the third year and planted to corn. In latitudes like that of northern Ohio, it is best to plant timothy in late September on top of the ground, never in the ground. Where winters are more severe, spring broadcasting gives better survival. Where timothy is planted with red clover or alfalfa, a rate of five pounds per acre is enough. If planted alone, eight to ten pounds is better.

I know of no grass grown for hay that is more abused in the making than timothy. It should be cut before the stalks head out, as I have said, when it makes a very high energy feed. That's why it was preferred for workhorses. It will then grow back quicker for a second cutting, too. But more often, either because of weather or because the farmer is looking for more tonnage rather than more quality, the timothy is allowed to head out, even beyond bloom stage, and the extra tonnage does not amount in food value to half that tonnage if the timothy were cut earlier. I have seen farm animals eating timothy that was little more than a bunch of brown sticks. They had to be half starved or were being fed a high-energy grain ration. Good timothy could have replaced both the sticks and the grain.

But cutting timothy after it is past its peak nutritional value has one nice side effect. As the mower moves through the stand, the seed heads in bloom puff little explosions of pollen as the mower blade clips off the stalks, a striking sight repeated zillions of times as one mows across the field.

Tall Fescue

Fescue is a controversial forage plant in grass farming. Even though graziers find faults with it, most of them still grow it. The first reason is that it can be grazed in winter better than other grasses. It is the key to year-round grazing. I thought for a while that was the end of its advantages. Now I think differently. *If it is clipped twice to keep it from going to seed and to keep it short and tender,* it holds up better than any other cool-season grass for grazing in hot, dry summer, too.

If you don't keep it short, you will probably curse fescue in the summer like I used to do. The livestock never prefer it to bluegrass or ryegrass, but they will eat it unless it gets long and wiry. In winter, they nose down through the snow to eat it. I'm told that fescue is rough and tough to the animal's tongue and mouth when it is growing fast in June, but cold weather tenderizes the plant. Clipping must do the same.

It grows and heads out very fast in spring. I try to mow it right before it goes to seed in May or June. The regrowth goes to seed more slowly, and I try to get it mowed again in July. The second regrowth does not usually seed and can be pastured and/or stockpiled for winter grazing. If rains are abundant, I clip it in August, too, just to keep it tender and to keep it from overwhelming white clover. It still grows back for winter pasture. Tall fescue forms an extremely tough sod, which is an advantage for winter grazing since cattle can walk on it without overly damaging the soil surface in muddy weather. The dense sod also blots out weeds effectively.

The first popular tall fescue was named Kentucky 31 and was hailed as the miracle plant of grass farming. Thousands upon thousands of acres were seeded to it, especially in the upper South. Then farmers discovered that Kentucky 31 was infected with endophyte fungi, which can cause "fescue foot" in cattle, pregnancy disorders in horses, reduced milk production in dairy cows, and other problems. Thousands of acres of fescue were sprayed with herbicides to get rid of it even though the bad effects were not consistent. And if 20–30 percent of the pasture was in other grasses or legumes, beef cattle and sheep could graze it without noticeable harmful effects. Bob Evans rates Kentucky 31 fescue at the top of the list of important grasses in his area. "Stockpiling Kentucky 31 fescue for winter grazing is an ideal starting place for farmers who wish to stay on the land," he writes to me. "We have thousands of acres of it at hand. I feel sure that if the farmers don't move in this direction, they won't be around long."

Endophyte-free tall fescues have now been developed. They are not as enduring as the old Kentucky 31. Especially in the hot, dry weather in the upper South, where tall fescue thrives, the stands tend to die out rather quickly. But plant breeders are working hard to overcome these problems because an enduring, endophyte-free fescue could be the key to practical year-round grazing even in the North.

One claim that university plant breeders made about Kentucky 31 was that it would not spread. That was the main reason I planted some in the first place, before I understood that experts can make dumb mistakes, too. Kentucky 31 does spread, believe me, and to this day I can't understand why the breeders made such an erroneous claim. I guess they meant that it won't spread by root stolons. But the plants sure do spread by seed. Road maintenance crews started using the grass to seed down embankments. The fescue then spread into nearby pasture fields, and some graziers, especially dairy farmers, were upset, to say the least.

But even endophyte-infected fescue is not all bad, and in many cases it is good. I have watched it grow and proliferate from a roadside embankment planting into my sister's pasture field for thirty years. Sheep or horses have grazed it all that time, even through the winter here in northern Ohio, without any bad health effects as far as any of us can tell.

My Kentucky 31 spread from the paddock where I planted it into two more, but after ten years, the first planting has thinned and bluegrass and white clover have infiltrated it because I keep it clipped. In fact, it is dying out in some places. We have grazed horses and milk cows as well as sheep on this fescue. The horse had her colt just fine, the cows showed no diminishment in milk production, and the sheep relished it for winter salads. One spring I burned the heavy mat of old fescue (I was not then winter grazing as much as now), and while I can't say for sure if the fire did it, that's when the bluegrass and clover seemed to get a foothold and slowly overwhelm the fescue. The combination of mostly bluegrass and white clover interspersed with some fescue seems to me the best nearly yearlong pasture that I have been able to establish.

Where the fescue spread into paddock B (see map in chapter 2) on its own, it is now the dominant forage, but it has never replaced all the bluegrass and white clover. I hope to keep the balance between them. If I ever want more fescue in the mix, I will simply not mow one time.

The main management step to avoid endophyte problems is to make sure that your livestock always have access to other grasses and legumes and not just infected fescue exclusively. I also think that fescue has more of a tendency to cause trouble in the South than in the North. Perhaps it is weather related or soil related, but there is definitely a Fescue Belt, as it is referred to, in the mid-South where endophyte problems are more prevalent. In that belt, it seems to me prudent for a commercial dairy or any commercial operation where rate of gain is crucial to profit to grow only endophyte-free fescue, even if it does not survive as well or is more susceptible to attacks from insects or disease, as it usually is. For the garden farmer or the part-time grazier concentrating on low-cost cow-calf or sheep production or where top rate of gain is not crucial to profits, infected fescue will work out for you just fine and be more permanent. Just make sure livestock has access to legumes and other grasses too.

Orchardgrass

Although commercial farmers in the lower Corn Belt and mid-South regions favor a combination of orchardgrass and alfalfa as their hay crop, I don't like orchardgrass and my sheep don't like it much either. It grows too fast in spring and easily gets ahead of the grazing animals or my haymaking schedule. I could never get it made before it went to seed and lost much of its food value. So I quit growing it. Because of its tall growth, orchardgrass for hay will overwhelm ladino and white clovers grown with it. Combine it with taller-growing alfalfa or red clover. Orchardgrass likes to grow in pronounced clumps, too, rather than spreading out to make a solid sod, a characteristic that for some reason irritates me. I don't like to see bare ground in a hay or pasture field. Late growth after being grazed or clipped twice is not any more palatable than tall fescue, nor is it nearly as enduring for winter pasture. Orchardgrass is not winter hardy in the northern Corn Belt.

Orchardgrass, however, grows better in hot midsummer than do other cool-season grasses except maybe bromegrass. That is the main reason farmers grow it in the midsections between North and South. It is less demanding of fertile soil, too, although, like any plant, it grows better on good soil.

As interest increases in grass farming, science is coming to the rescue with improved varieties of orchardgrass that do not mature so quickly in spring and supposedly have better palatability. On your soil you may very well find a variety or strain that will serve you better than improved varieties of timothy. But don't bet on it. At the last meeting of graziers I attended, a show of hands suggested that orchardgrass is decreasing in popularity.

It is best to sow orchardgrass in the spring. It can be broadcast on top of bare or partially bare soil, as described for other grasses and legume. Sow three to five pounds per acre in combination with a legume.

Smooth Bromegrass

Of the many types of bromegrass, this one is of most importance for grass farming because of its wide adaptability. It grows all over the northern half of the nation, where it is most often combined with alfalfa. Bromegrass's salient advantage is drought resistance, and since that is true of alfalfa also, they go together like bread and butter. But brome is still a cool-season perennial and will go dormant if the weather gets really dry. And I don't think it competes with alfalfa as well as orchardgrass. According to *Forages*, it is more important with brome than with other grasses discussed here to head off dormancy by making the first cutting of hay or by grazing before it matures and starts heading out.

I used to try to grow bromegrass. I found it hard to get established in my soil and quit trying.

Questionable Grasses

Quackgrass is a fairly good pasture grass, no doubt about it. So are crabgrass, yellow millet, and foxtail. But, except for millet, they won't compete with sods of bluegrass, ryegrass, and clover. They like cultivated ground. That's why they are a problem in grain farming and other crops. I wonder if I am wise even to mention them. On the other hand, I can't resist, because if the tools of soil disturbance were put away for good, these grasses would not be noxious weeds and would be kept under control by the chomping teeth of grazing livestock. I like to see millet in the pastures. It is more a warm-season grass, and for a brief time in high summer it shows a lot of green before it heads out all too quickly.

Pestiferous grasses provoke a philosophical attitude about life. Almost invariably they become pests because they like disturbed soil. If we didn't cultivate, they wouldn't be pests. In Minnesota I more than once dug potatoes that were pierced through and through by the roots of quackgrass. I have also observed quackgrass roots inching into the edge of a road and heaving up the blacktop. I helped my father-in-law hoe patches of Johnsongrass out of his cornfield; two weeks later you could not tell where we had hoed. I have seen large fields in Kentucky abandoned to six-foot-tall Johnsongrass and torrents of woebegone cusswords. The only way to control it is to graze it. In fact, it was introduced as a good grass for grazing and, managed closely, can be just that. But even then care must be taken. It is toxic at certain stages of growth. Many farmers in the South hate Johnsongrass so much they will remove their shoes upon leaving a field where Johnsongrass grows to make sure that not one seed clings to their socks or shoestring eyelets. But they keep on cultivating.

A farm near Mt. Vernon, Ohio, years ago became so infested with foxtail that the organic farmer operating it gave up trying to control it and made hay from it instead. Turned out to be excellent hay. But in a few years it declined. It needed the soil disturbed by cultivation periodically in order to thrive.

Crabgrass is the bane of lawns, but Red River crabgrass is much praised by some graziers. You can find out all about this grass and the others mentioned in the literature of pasture farming, but I am not going to help you out because, all things considered, I think in most cases there are better alternatives among the other grasses. And I am sure that the garden farmer and part-time grazier are better off to avoid them.

The same thing is true for the sorghum/sudan grasses that professional graziers grow for high-production pastures in midsummer. These grasses are slightly toxic when young. The general rule is to wait until they are over a foot tall before grazing.

I had experience with reed canary grass when I was in Minnesota. Many farms there had low marsh holes of an acre up to several acres in size between hills. These holes were mostly untillable. Farmers planted them to reed canary grass for grazing or, on the solider portions, for hay. Reed canary grows tall and rank and not very nice hay, but it will provide pasture in the hottest part of summer. It can put down such a mat of sod on marshy soil as to hold up a small tractor where it might otherwise sink to the hubs. I think the best thing you can do with those marshy wet holes where reed canary likes to grow is turn them into wetlands.

Because of its matted sod, reed canary grass is a good grass for erosion control in gullies. But, as this description tells you if you read between the lines, reed canary grass is a grass for problem soils, and problem soils do not make very good pasture farms. In other words, a farm relying on reed canary grass as a pasture crop is a poor farm indeed. Reed canary is also slightly toxic.

Strictly Southern Grasses

Bermudagrass always reminds me of my mother-in-law, who hated the stuff when it got into her Kentucky garden. Even into advanced age, she was out there in the blazing sun battling bermudagrass, her face red with heat and anger enough to cause a stroke in a lesser human. Bermudagrass may be a blessing in the Deep South, where few grasses will survive as well as it does, but in the rich soil of my mother-in-law's garden, it was an unkillable plague. To keep the root stolons from invading her tomatoes, she dug a trench around the garden and inserted a wall of plastic film. The bermudagrass usually found a way through, under or around the plastic, but she never admitted defeat.

I suppose that's as high a compliment as I can give to bermudagrass for those thinking to grow it as a pasture in the South. Just don't put your garden anywhere close. Starting a pasture of bermudagrass should not be a problem. Put one plant in a ten-acre field and run for your life lest it catch up and strangle you. In the early days of bermuda-grass, when southern farmers were more interested in cotton and corn than yearlong pastures, they were in total agreement with my mother-in-law. They tried every possible way to kill the stuff. Some states passed laws prohibiting its introduction.

But bermudagrass has become for the South what bluegrass is for the North, the old reliable for grazing. Coastal bermudagrass does best in the Deep South along the ocean coasts, as the name implies. Mid-land strains are grown in a narrow midsection of the upper South, with a northern boundary of about the Ohio River. These hybrids do not yield as well as coastal bermudagrass but survive colder tempera-tures.

Dallisgrass is a good southern pasture grass It begins to grow early, even for the South, and grows later in the year than bermuda-grass. It has, in fact, the longest growing season of all southern grasses, nearly year-round, and so is very desirable in pastures. It grows in the same climate as cotton, from east Texas to the Carolinas. It is more tolerant of wet or dry conditions than bermudagrass. It grows in bunches, so there is room between the bunches for legumes to get es-tablished. Late winter to early spring is the preferred seeding time. A pound and a half of seed per acre is recommended. I notice that *Forages* suggests ten pure (viable) seeds per square foot as a seeding guide. I like that kind of guide. Mechanical seeders, especially broadcasters, are not very accurate, and I set the gauge not by the numbers given on them but by the amount of seed being spread. (For the legumes and grasses I sow, I like to see at least one seed on the ground for every square inch.)

Native Prairie and Plains Grasses

The native wheatgrasses, bluestems, gramagrasses, and switchgrasses suggest cowboy country, the vast ranges where once buffalo roamed. It is here that many of the ideas being adopted by grass farming in the other two-thirds of the nation have been practiced the longest and need to be studied by us "Easterners." For example, the goal of year-round grazing has been sought longer and been more nearly achieved in parts of Wyoming than in parts of Ohio. Using introduced crested wheatgrasses, pubescent wheatgrass, introduced Russian wild ryes, and Regar meadow brome, cattlemen and sheep producers on the northern plains can graze through the winter. These species stand up well to snow cover and preserve enough of their summer nutritional value that in some cases supplemental hay is not needed. Sometimes these grasses are cut and windrowed for winter grazing.

Some of these native grasses or their close relatives were also native to the prairies of the Midwest and are being looked at carefully for possible application to grass farming everywhere, especially for late summer grazing. At one time these grasses covered much of what is now Corn Belt country, growing so tall that early settlers even on horseback could get lost in them. Their advantage is that as warm-season grasses they flourish in midsummer, when cool-season grasses slow down. Switchgrass, big bluestem, indiangrass, and eastern gamagrass are drawing the most attention. They are tall, up to six feet. Sometimes, in the case of eastern gamagrass, they are twelve feet high and grow in bunches. Although I have not found these native grasses any better for summer drought grazing than alfalfa and red clover, they are perennial and, once established and managed so that they have proper time to regrow between grazings, are more or less permanent. One advantage of these native grasses is that they can be safely grazed after frost. Some warm-season grasses, especially in the sudan family, increase enough in prussic acid content to be toxic after frost. Like most other plants, the native grasses should be allowed to store energy in the roots from September to frost (August to frost in the case of eastern gamagrass) and during that period not grazed. After frost, they should be grazed down to about an eight-inch stubble to go into winter.

I bought some switchgrass, bluestem, and eastern gamagrass seed once (very expensive), the three native prairie grasses best adapted to Ohio. I learned that the cautions from pasture scientists were correct. These grasses are difficult to get established. I lacked the patience needed (two years) and could not keep weeds from overwhelming the native grasses. Also, I was learning that for summer grazing legumes that I knew how to grow worked fine and were more palatable. I just couldn't make these tall bunchgrasses work for me. Why grow eastern gamagrass if I have to quit grazing it in August, right when I need summer grass the most? Also, because of the grasses' big size and heavy forage yields, graziers use fairly heavy fertilizer applications or interseed red clover or lespedeza with them for nitrogen. That can be difficult, because the warm-season grasses grow so tall that they block sunlight to the legumes. Moreover, if you can grow good red clover or lespedeza, which supply good forage in late summer, why mess around with the warm-season grasses?

I worry that my lack of enthusiasm about tall bunchgrasses may be a mistake over the long haul. I am no doubt overly influenced by my failure to establish a stand. Bob Evans and his farm manager, Ed Vollborn, have had surprising (to me) successes. Graziers who are looking for summer pastures that, once established, will continue indefinitely without reseeding will not be as negative as I am. It may even come to pass that pastures turned back to native grasses for a long time, like a hundred years, might produce more forage and more economical meat production than the Corn Belt ever dreamed possible with corn. At one time, without any management or husbandry at all, unnumbered millions of buffalo thrived on these grasses.

Future graziers from Iowa to New York will surely try switchgrass, which seems the best native warm-season grass for the Midwest and East. It grows to five feet tall and adapts to a wider range of soils than the other native grasses. The Cave-in-Rock variety is considered the most palatable. Switchgrass seed is free-flowing and can be seeded with standard seeders, not necessarily true of other native grasses. Sow six to ten pounds per acre. The other native grasses mentioned have chaffy seed that is difficult to sow without a special grassland drill such as the Truax.

An aside that I can't resist: In researching warm-season grasses, I happened to run across in an article (I'll not say which farm magazine it appeared in or which university released it) this sentence: "Yield of warm-season grasses will depend on soil fertility, dryness of soil, rainfall during the growing season, and the amount of fertilizer used." Well, bless my soul. This kind of vacuous sentence is common in farm and garden publications. I'm afraid I've been guilty of it myself. The only reason agricultural journalism can get away with it is custom, I think. Farm writers I know are not stupid. We waste space to say the obvious because of careless habit or from reading too many seed catalogs. I have heard farm publishers say that farmers have a reading attention span of only three minutes. No wonder.

As for the true rangelands of the Plains and the West, a subject somewhat beyond the scope of this book, graziers everywhere should read Henry Turney's *Texas Range and Pastures—The Natural Way*. Mr. Turney says he has not had to feed one bale of hay to his cattle since

1947. He says most of the problems of cattle country come from over-grazing. He advocates rotating pastures so as "not to allow the grass to be bit off as soon as it would stick its head out of the ground." He goes so far as to advocate giving a part of one's pastureland a complete rest during the summer at least once every three years—just not grazing it at all for several months. That of course means moderate stocking rates, which flies in the face of the cattleman's natural desire to make more money. Says Turney in an interview in the April 2003 issue of *Acres U.S.A.*: "we seem to be people of numbers. Maybe it is human nature. When rangemen around here visited, their main question was 'How many cattle do you run?' They seemed to think, and still do to a large degree, that the number of cattle you own is correlative to the amount of income you make. We know differently of course. We know it depends on how fast the calves grow and how much you have to feed the cows to raise the calves . . . it is more economical to have your land feeding your cows instead of you buying their food." He also says, "I have used, in some cases even recommended, some of the introduced grasses, but I don't know of any introduced grass that can compete with the natural grasses." He was speaking mainly against introduced varieties like "old world" bluestem and for little bluestem, the native grass he cherishes and which, he points out, is mostly gone because of overgrazing. Something else he says in the interview is worth repeating: "I was out here minding my own business, and the son of a good friend of mine, who was with Tarleton State University in Stephenville, Texas, came by my place and wanted to know if I would teach over at Tarleton. I said no, I'm a rancher and a farmer. Well, he kept asking if I couldn't take time to teach a class and finally I agreed to teach a course in range management. Lo and behold—and I get mad every time I think about this—there wasn't a decent text that I could find. All of those PhDs who had been teachers all their lives at A&M and Texas Tech and those places, and they didn't have a book. I don't know how they taught. That is how I came to write my book."

17

Grains for Grazing

Although graziers in the South and southern Plains have for years grazed their livestock on oats and wheat, the practice until recently fell into disuse in the North when it looked like growing grains for the cash market was the wave of the future. Cash grain seemed to make money, and did for awhile, helped along by subsidies, because grains grow so well in the Corn Belt. But that is also why there is such a great potential in the same area for growing grains for grazing. I speak now not only of grazing grains as grass, which is what is commonly done by cattlemen in the South, but also of grazing grains for the grain. Grazing corn with pigs was common in earlier years, but doing it with cereal grains is a fairly new idea. Grain-grazing appears to circumvent the commonest criticism of grass farming, that grass and clover do not provide enough concentrated energy to fatten meat animals or maintain milk flow enough to compete in the mainstream market. But if animals can graze high-energy grains in the field along with the grass and clover, they could get all the "concentrates" they need (or the market thinks they need) but not *more* than they need, which is what often happens in the feedlot and milking parlor.

Is it practical to graze grains? What follows is some evidence that the answer is yes.

Oats

Perhaps what will be hailed someday as one of the most significant experiments in twenty-first-century farming took place last winter (2002–3) on the Wolfinger farm near Lancaster, Ohio. In what is already being referred to as "the most famous oats field in recent Ohio history" (*Farm and Dairy*, June 19, 2003), the Wolfingers no-tilled oats on August 5 into a thirty-acre field that had been in wheat previously. On November 8 they began grazing the oats, which was about thirty inches tall and headed out, with forty-four head of beef cows and a bull. The yield of the oats in dry matter was estimated at five tons per acre. The protein content was almost 20 percent. The cattle grazed the oats in strips about 8 feet wide and 950 feet long every day and were fed in addition one round bale of low-quality hay per week. In mid-February, grazing had to stop temporarily because of a snow and ice storm that ended with a blanket of ice over the snow. I would have thought that no forage except corn would have been grazable after that ice, but in this experiment, the cows were back on the oats on March 6 and continued to graze for three more weeks—until the oats was all eaten up. The calculated *net* return was over $100 per acre, all from a field that, as the article I am quoting from remarked, *"never saw a combine."* It also never saw a plow, or a disk, or a field cultivator, or a spray rig, or a grain transport wagon, or a dryer, or a grain bin in producing that oat crop.

Anyone having a hard time believing that experiment should observe how the Amish treat their gardens in winter. They sow oats heavily in the fall, and the garden remains covered with a lush stand of it until January. The crop continues to be good forage even as it turns brown into March. Since oats dies over winter and does not grow back like wheat or rye, the gardener can more easily incorporate the plant residue in the soil in spring. Oats carries some natural herbicidal power, too, and no doubt suppresses early weed growth when planted so thickly.

I had to try to replicate the Wolfinger experiment. I had already done it in the garden, and I could hardly wait to try it in a field. But I added another wrinkle to the experiment that partially negated its success. Because of all the rain in 2003, I did not get my little field of corn planted until June 8, and then only half of the corn came up. So I decided to interseed oats into the corn, hoping that enough sunlight would filter through to grow the oats. I broadcasted the oats by walking down between the corn rows, cranking on my little seeder. Then with a garden tiller I went down the rows again, lightly covering every other row middle. I have never been able to get as good a stand of oats by broadcasting on top of the soil as I do with legumes, so I wanted to compare surface-sown oats with oats lightly covered with soil.

The oats did come up better where it had some soil covering, and so I have to conclude that oat grains do not germinate on top of the soil as readily as legume seeds. At any rate, the stand was adequate overall. The red clover I sowed after the light rototilling was even better.

Both oats and clover grew well in the light shade of the thin-planted corn for the first month. Then the oats, at about ten inches tall, blighted and quit growing. The corn shade was still too much for it, I presume. The red clover, however, continued to grow very well. Later, the oats recovered, but by then it did not head out as well as it would have otherwise.

I write this in September as I ready the book for publication, wondering if the late-planted corn will mature. Even if the oats doesn't amount to much, it is obvious that the red clover is going to make great winter pasture and a full season of hay or pasture next year. If the corn doesn't mature, it will still make grazing grain along with the fodder.

Ed Vollborn, the retired Ohio State University extension specialist mentioned previously, said to me recently that oats shows much more potential for grazing than has yet been realized. His way of putting it was: "Oats goes from zero to sixty faster than any other potential pasture plant." He meant that within two months after planting, oats can be grazed or hayed. Planted in late April, as we normally do in Ohio, it can make pasture by late June. Allowed to mature, it makes grain by the middle of July or sooner. Now I also understood that oats planted in August is ready to graze by October.

There are other uses of oats sown at the usual time in spring. It is difficult to make hay from it because the stems are slow to dry, but livestock love oat hay. Sometimes I try to cut mine for hay before it heads out, when the stems are not yet so thick. Then it will grow back and make a second, shorter crop, which I may cut for hay again but more likely will graze in late July or August, when permanent pastures are in decline. The seed heads at that time are in the late milky stage or are beginning to mature, and they make an excellent grain supplement for meat calves and lambs, lactating mother ewes and cows, and gestating sows and their litters. Since I always sow red clover or alfalfa with the oats, the ripening grain has an underlay of lush legume to make a total feed ration.

If I wait until the grains are mature to turn the animals into the first or second stand of oats, something wonderful happens. The grazing animals will knock to the ground some of the oat grains, and when rain falls in August or September, a new stand of oats grows for fall or winter grazing without any help from me. It is not as strong a stand as what I interseeded with the corn in August, but it does add some grain to the legume for fall pasturing.

It is practical to make hay out of mature oats, too, although many farmers, growing up as I did, will think it odd to do so. I learned the practicality of doing this from Karl Kuerner, a farmer in Pennsylvania who became very famous because he was the subject of many of Andrew Wyeth's renowned paintings. Karl was one of the most frugal farmers I have ever been blessed to know. Facing the prospect of having to buy a grain harvester for his oat crop, which wasn't really large enough to justify that kind of investment, he simply treated the oats as if it were hay. He already had hay equipment. He cut and baled the oats when it was ripe, then fed the bales to his animals through the winter. They ate the grain and some of the straw, and what they did not eat became bedding. Imagine my chagrin when years after I learned about Karl's cleverness, and praised it in print and practiced it myself, I read in an old (1948) USDA Yearbook of Agriculture that "nearly as many acres of oats are cut when approaching maturity and fed unthreshed as are cut when less mature and cured as hay."

Since I can't justify a baler on my small farm, I pick up the windrowed oats with my hay scoop on the tractor and stack it in the field for the livestock to self-feed from in winter. I must be sure, of course, that the oat heads get dry enough in the windrow to be stacked without molding. Oat straw in stacks has an advantage over legume haystacks in that sheep can burrow into the stack while eating without getting as much chaff in their neck wool. The stack may draw mice over winter, but not to a degree that will cause a problem, as might happen if the oats were stored in the barn.

Feeding oat straw even without the grain heads in it is an old established practice. If you split open a yellowed oat straw with your fingernails, the inside stem is often surprisingly greenish. Animals grazing lush legumes or eating lush legume hay will readily eat oat straw for roughage. The extra fiber is just what they need to counteract the bowel looseness that lush legume hay (or silage) may cause. The farmer I worked for in Minnesota when I was young told me that after an especially bad summer drought, he kept his milk cows alive all winter on his big stack of oat straw. Although the straw stack was meant primarily for bedding, it was the only feed he had. He said he didn't get much milk that winter, but no cow starved, either.

Traditionally, oats is sown in the spring on fields that have been in corn the previous year. In my continuing effort to avoid machine and fuel costs, I frost seed it very early in spring when the soil is still freezing and thawing. Broadcasting it this way does not make as good a stand as when I sow on disked land in late April or lightly cover it in an August planting, as described above. But broadcasting can make an adequate enough stand without disking if I sow the oats late in winter at double the usual rate of two bushels per acre. As long as the weather is cold, grains will not sprout, and so they lie unharmed on the soil surface while constantly coming into better contact with the soil from snow, rain, freezing, and thawing. When warm weather does come, they sprout and grow. We have for years taken advantage of the ability of seeds to stand winter weather and start growing in spring, often ahead of weeds in the garden. For example, we don't have to plant lettuce. Seeds of lettuce fall on the mulchy soil surface in the autumn, and in spring they volunteer on their own and are ready to eat before lettuce planted in the cold frame.

I always sow clover or alfalfa directly following the oats. If I broadcast on disked soil, I first sow the oats at two bushels per acre and then go over the field with the cultipacker, aiming to get the oats pressed solidly against the soil (what frost seeding does without a cultipacker). Some seeds go maybe an inch into the ground and some remain on top of the ground. Both usually sprout and grow okay. Then I sow the legume seed on top of the ground. I would like to buy oats from a farmer, since the market price for oats is only around $1.70 a bushel most of the time, but it is difficult to find on-farm oats in our area. Few farmers grow it because of the low price. I have to go to the elevator and buy shipped-in oats at five times the on-farm price. If you raise horses and buy a lot of this high-priced oats, sit down and figure out how much you could save if you grew oats for your horses as a pasture crop.

Wheat

Sowing wheat by broadcasting on the soil surface usually makes an adequate stand. Many grain farmers broadcast their wheat from airplanes and get adequate results if they double the normal seeding rate, from two to four bushels per acre. Grain farmers also effectively use no-till drills to plant wheat into soybean stubble or other plant residue from a previous crop.

I sow wheat in paddocks where legumes are declining. In an effort to avoid using machinery and fuel, I like to turn livestock into the legume in August and September and let them overgraze and trample the soil into partial bareness. This is called "mob-stocking" by professional graziers. With the paddock mostly bare and weeds not much of a factor that late in the year, wheat sown on it will sprout adequately for grazing, but usually not heavily enough to make a profitable stand for harvesting as grain. The more livestock walk over the paddock or the more animals you concentrate there for a short time, the more the seed gets pressed firmly into the soil surface by their hooves, and the better the subsequent stand of wheat. When the wheat, sown in mid-September, has attained a height of four inches or more, by about November 10, when growth ceases, I again turn the livestock in to eat down the stand. The first time I did this, I feared it might cause the wheat to suffer winter injury, but that has not proven true. Weather that causes winter injury to wheat will do so whether the stand is grazed in late fall or not.

In late winter I broadcast a legume and timothy into the wheat if I am taking the paddock back into a hay crop. If I am going to plant corn, I simply let the animals graze the wheat hard in spring until corn planting time. If corn was not in the picture, I could graze the wheat a little in April, like the Kansas wheat growers do, discontinue grazing by early May, and let the wheat grow back for a grain harvest or, in my case, for more pasture. Being inclined from my upbringing to think of wheat as a cash grain crop, I have been surprised at its resilience as pasture. It can be grazed quite severely in May and will just keep on coming back. This year, after heavy grazing, I did the first disking in preparation for corn. Rain came before I could get to a second disking and planting. The wheat, though harmed considerably by that first disking, rerooted itself partially during three dismal weeks of rain and kept on providing pasture while I waited for dry planting weather.

When the wheat is a nurse crop for legumes, spring grazing will not injure the tiny legumes getting started in it—unless grazing is overdone. I learned this first from Steve Gamby, a forward-looking commercial dairyman not far away. He has utilized fields of wheat entirely for grazing—that is, with no subsequent grain harvest. The seedling alfalfa in the wheat is not only unharmed by grazing down the wheat but, it seemed to me, helped by it.

I have not tried to make "hay" out of ripe wheat as I have done with oats. I suppose it would work, but wheat straw is not as palatable as oat straw. In experiments, I did feed ripe wheat on the stalk to my chickens, and that worked fine. Chickens will also "graze" ripe wheat off the stalk in the field. I can think of no grazed grain that would be as effective a source of energy and protein for chickens than a ripening wheat field. Chickens don't much like oat grains because of the tight hull around the groat, but they love whole wheat. Wheat contains nearly as much energy, pound for pound, as corn, and the grain heads on the wheat stalks are easily enough reached by grazing chickens. Livestock grazing ripening wheat would also be practical, I think, but I have not tried it. A neighbor did once, but not deliberately. His cattle broke through the fence. They seemed gloriously happy munching the wheat heads. Not having tried it, I am not sure if hogs would graze wheat as well as they do oats. An old traditional belief states that care must be taken in feeding whole wheat to hogs. Hogs will swallow too much whole grain without chewing it, so folklore says, and the grains can swell up in their stomachs, causing problems. I doubt this and have fed small amounts of whole wheat to my hogs with no ill effects. F. B. Morrison's *Feeds and Feeding* doesn't mention any such problem feeding wheat to hogs but instead says that wheat makes good hog feed. Perhaps a starving hog suddenly having access to a whole bucket of pure wheat would gulp it down rapidly and suffer digestive problems, but no good husbandman is going to allow that situation to occur. Hogs slowly grazing off grain heads with straw and chaff are not going to have a problem.

Using wheat as the transition crop between a declining legume stand and reseeding a new one, instead of growing corn in between, I do not need to cultivate at all, as I have described. The traditional farmer in me gets very antsy when the wheat does not make the beautiful dense fall stand that cash grain farmers need to make a profit. I have to learn patience. Eventually, the wheat thickens up well enough for profitable grazing, and it is adequate for harvesting a little grain, too.

Corn

In former times, corn was often grazed, by hogs first and cattle afterward, but thinking of it as a grass to be grazed is a fairly new idea. Corn, so far, has not seemed to fit integrally into pasture farming, mostly because it is a one-shot crop. It takes most of the summer for it to mature. You can't graze it repeatedly, and so you can't control weeds except by putting it in rows and cultivating between the rows or by using herbicides. I once observed a field of corn that by some oversight was neither weed-cultivated nor sprayed for weeds. The giant ragweed totally overwhelmed it.

However, once corn has grown to its full height, it can be grazed. I have already described how, but it can stand repeating. First, lambs are put in to eat weeds that grew after cultivation and to eat the lower leaves of the cornstalks. Lambs generally don't bother the ears, especially if the corn is a tall, open-pollinated variety where the ears form high on the stalks. My grandfather and, in later years, my good neighbor, Bob Frey, followed this practice. Then they "hogged off" the corn—that is, they let the hogs in to eat the ears. In the winter the cows and sheep are put in to eat the corn fodder and whatever ears the hogs missed.

Since cultivation, either for a seedbed or for controlling weeds, seems to me integral to corn production, it doesn't fit into my ideal of a till-less pasture farm. It could, however, with a no-till drill and herbicides. Since neither of these practices fits my little acre of corn, the only way I can justify it as a tilled crop is that it occupies only a small portion of the farm every year in a five-year rotation with forage crops. Even then, the real reason I do this is to experiment toward the time I can learn how to winter pasture the corn, ears and fodder both. The more palatable fodder and the ears remain above the heaviest snow.

Corn can be broadcast planted; farmers did it in the 1950s with airplanes. It is just not very practical. Corn kernels left in the field after harvest will even survive winter and grow in the spring. Then they become a "weed" problem in the following crop of soybeans. One of the amusing developments in grain farming (not amusing to grain farmers) is that when corn with the gene making it immune to Roundup herbicide volunteers in the soybeans, which are also immune to Roundup, the herbicide can't kill the corn.

Grazing corn when it is still green is on the increase. Varieties are being developed specifically for grazing, especially during the late summer decline of cool-season grasses. Again, I am not going to mention named varieties because that information could be obsolete tomorrow in the fast-moving world of grazing. Consult grazier seed suppliers. Corn for cows and beef cattle is usually strip-grazed with a single electric wire. This usually means knocking down an alley of corn to run the wire through, but the animals will graze the fallen stalks just as well.

My sheep and cows have always loved sweet corn fodder; they will eat a sweet corn stalk right down into the ground. On regular field corn, animals usually don't eat the lower stalks. So, this year I planted a large patch of sweet corn next to the field corn, specifically for grazing. I used a special strain of Delectable, which the local seed company, Walton's Seed, sells. We like it the best of all sweet corns, so I figure the animals will too. Planted late, it grew only about four feet tall, so (and this is very important) the red clover I interseeded in it is growing fine. I did not interseed oats in the sweet corn, but I should have. Amazingly, the raccoons have not decimated the crop, and I know the livestock will make short work of the fodder and the ears we don't eat. Plus I will have a new crop of red clover next year to graze or make into hay in June.

I also go on growing corn because I don't think it is possible to convince farmers, including myself, to stop completely. Rather than wasting time trying to convince us that corn is not necessarily necessary to animal food production, or that it would be better for most farmers to buy it than to grow it at today's cheap prices, I would rather convince farmers to change their present mix of 90 percent cultivated crops and 10 percent pasture to something like 80 percent pasture and 20 percent corn. I think that might turn out to be more profitable than pure pasture farming or pure cultivated crop farming.

Growing corn is fun (which I think is the real reason we keep on doing it) if you don't mind losing money or don't mind standing in the FSA office for hours on end to make sure you get enough subsidy to cover your losses. A farmer friend of mine, Dave Frey, who is also sort of a genius, says that there is no experience more pleasurable than cruising down the corn rows in a $200,000 combine with what seems like unlimited power at the touch of one's fingertips, complete with phone, radio, CD console, and air-conditioner (refrigerator is an extra), watching the golden corn kernels pour into the bin—*even when you know you are losing money with every kernel.* Another acquaintance, Herman

Schoenburger, who truly was a genius as an electrician (he's passed away now), told me once that after years of cutting, shocking, and husking corn by hand, and then by early machines, the first time he put a picker-sheller corn combine through its paces, he literally broke down. "I cried not only because it was so beautiful to harvest corn this way, but from remembering all those years of pain and sweat, harvesting it the old way." I wonder if I can use that remark to convince farmers to make corn a minor crop and let the livestock do the work.

Corn is also my cover-your-ass crop. I aim to prove that it is unnecessary and even unprofitable on most farms, but just in case I'm wrong, I keep on growing a little. Like an old farmer in Minnesota, Ed Kraus, once said to me, mimicking Blaise Pascal without knowing it: "I don't really believe in God, but I keep going to church just in case I'm wrong." Working a paddock of corn into my five-year rotation of temporary pastures is actually more trouble than it's worth, but I defend my stubbornness by pointing out that as a portion of the whole farm dedicated to pasture farming, only something less than one-tenth of the acreage is in corn in any given year. And always on level, not hilly, land. If that corn eventually becomes a way to ensure practical winter grazing, then I figure I've made the right decision. If it doesn't, I have the corn for supplemental feed for hogs and chickens and an occasional treat for the cows and sheep. Actually, I think of corn now as my sheepdog. When I jostle a few ears in a bucket, the sheep recognize the sound and come running.

Since on my small pasture farm I grow only about an acre of corn per year (to yield at least one hundred bushels, though twice that amount is within practical possibility), I do the planting and harvesting by hand, which used to tempt Herman Schoenburger to shed more tears. But at my scale, growing corn is recreation, my golf game. I use two garden push planters bolted together to plant two rows at once. The farmers in the bordering fields stare in amusement from their monster tractors and planters. They are always very polite when I talk to them, and I will never know, I suppose, if they think I'm insane. I could plant ten acres of corn every year easily enough with my hand-powered planter, which in my frame of reference would mean all the corn necessary on a hundred-acre farm.

If I did not have my own open-pollinated corn with its softer kernels, I would plant one of the new hybrids, such as Masters Choice, which is bred of softer corn specifically for silage. I feed my corn mostly to pigs and chickens, but I have fed whole ears to cattle and sheep. I was surprised to learn recently that even some farmers don't think cows will eat whole ears of corn. Sometimes I break the ears in two if they are especially large. The cob is good feed for cows, too.

I plant in thirty-four-inch rows (wide enough to get the rotary tiller between the rows for weed cultivation). I like to get the kernels a foot apart, though a little more or less is okay. I don't know yet what is the best spacing or plant population for my corn and soil fertility. But because of my new experiment with interseeding oats and clover, I plan to plant only half the kernels I normally do from now on so the corn won't shade out the interseeded crops. If I get two good-sized ears on every stalk eventually, I will not have lost very much corn yield. I till the weeds twice, and when the soil is loose from tilling, I go down each row, straddling it, and with my feet I roll in the dirt to cover the weeds in the row. The whole family has learned how to do this. We call it shuffling. Once you catch on to it, the job is fairly fast for hand work, or foot work. But if you let the weeds in the row get more than two inches tall, then shuffling to cover them becomes muscle-straining and sloppy work. You can roll in dirt nicely to cover small weeds with tractor cultivators, too.

At harvest time we husk the corn by hand, going down the rows and tossing the ears into the pickup truck. I keep getting smarter about this. I used to laboriously husk out every ear, including the nubbins, which are hard and slow to husk. Now I leave the nubbins and small ears for the sheep and cows, which I will turn in for fall and winter grazing. It makes much more sense, and it is a prelude to the days when I hope to winter graze all the ears.

Our children and now our grandchildren help, and the harvest has become an enjoyable family affair. I keep wondering if our grandsons will be the last generation ever to harvest corn by hand, or if they will be the first of a new era where garden farmers will do it as a matter of course. Since the job on a small acreage is more fun than work, I am just crazy enough to believe it is the wave of the future.

One reason the work is so much fun is that we have been growing our own selection of open-pollinated corn for more than twenty years now. Each year we save the biggest ears (actually, ears that weigh a pound or more rather than just the longest ears). As we hand harvest, there is a lively competition over who will find the grand champion ear for the year. My dream is an acre of corn producing 24,000 ears on 12,000 stalks, with each ear holding a pound of grain. That would amount to a yield right around 400 bushels per acre, something that has only rarely been achieved. We are up to about 200 ears a year that sport a pound of kernels, and each and every year the count goes up. If I could live to be four thousand years old, I just might do it. Joking aside, if forty generations of garden farmers keep up the project, who knows what could happen. The earliest corn, before the Aztecs got interested, was no bigger than a head of wheat. Every farm could then have its own corn, especially suited to its soil, not corn owned by monopolistic international megacompanies. I suppose one of those companies would try to patent our open-pollinated corn and then say we can only grow it if we pay them a fee. In a pig's eye. That's the kind of thing that is going on, you know: patent protection as a cover for stealing plant germ plasm from the centuries-long cooperative work of man and nature.

Barley

Barley is grown the way wheat is grown, sown in the fall here for next summer harvest. It can be sown in the spring, too (or all winter long, as I have pointed out), in areas of severe winters. If fall sown, it is generally earlier to ripen in June than winter wheat. It can be grazed the same way as wheat, and now that the rough-awned (bearded) varieties have been more or less superseded by smooth-awned, or beardless, varieties, they don't cause sore mouths in grazing cattle. Except where barley is particularly adapted to soil and climate conditions, it probably won't become integral to pasture farming.

Lots of barley is grown in Scotland, where the climate is perfect for it. Scotch barley (and pure Scotch Highland water) is the basis for Scotch whisky. (The Scots spell it whisky, not whiskey.) For hundreds of years, and it still is true in some instances, Scotch liquors were the product of rather small farms, the perfect value-added grain crop. After the malting process and the distillation of the whisky, the grain was fed to livestock.

Needless to say, barley is also used to make beer, especially in the United States. As beer brewing, unlike whisky distillation, is being encouraged as a small local business, barley growers are wondering if they can't do what the Scottish farmers did: make their own barley malt and sell their own labeled beers. Quality grain alcohols for beverages could still be the top profit maker on American farms, as they started out to be, but our government, always looking out for the rich, took the right to make a good living from whisky and gin away from hill farmers and gave it to big business. So did Russia with vodka. Scotland still has a surprising number of small distilleries, but they are now almost all owned by big companies. Whenever a venture shows profit potential, it will one way or another become the domain of big, consolidated businesses.

But, I can entertain fantasies, and one of mine is a nation of pasture farms where part of the pastures grow grain for making good local liquors. Or, if that offends you, for making good local breads.

Rye

The popularity of ryegrass has diminished pasture farming's interest in rye. But rye still makes fall and spring pasture worthy of trying in cold northern climates where ryegrass won't endure. Cereal rye is, however, an annual and so it won't make a permanent pasture. The problem I see with rye is also its unproblem. Fall-sown rye grows early and fast in spring and can get out of hand as a cover crop. For grazing, you need to get animals on it as early in spring as possible to keep it from shooting sky high (six feet tall) in May, when it becomes less palatable. Even when rye is kept down by grazing or mowing, it is difficult to get another crop seeded into its tangle of roots without some serious soil cultivation. Since my goal is to cut down on cultivation expense as much as possible, I leave rye to hardier souls with heavy equipment. Otherwise, rye is one of the easiest grasses to sow by broadcasting. Even before the advent of pasture farming as such, it was commonly sown in cornfields in the fall to control winter erosion.

Triticale, Spelt, and Other Minor Grains

Arthur Jenkins, the fabled and long-lived editor of *Farm Journal* magazine in the first half of the twentieth century, liked to tell me (he was in his nineties then and still coming to the office) that there were no new stories, just new editors. There are no new crop plants either, just new farmers. Triticale was "new" in 1967, and I remember gushing in print about this "miracle plant" that was going to become more important than wheat and make farmers rich. Triticale is a cross between wheat and rye, supposedly combining the yield potential of wheat with the cold weather survivability of rye. I still run into new farmers and a few old ones who are enthusiastic about this "miracle plant." It has been grown successfully, for sure, but not often enough. Its popularity comes and goes. For all the editorial pomp and ceremony that attends its comings, its goings generally follow without fanfare. It just hasn't lived up to all our hymns of praise. Wheat continues to outyield it and, like rye, triticale is not as palatable to animals as wheat.

Interest in spelt, a mostly forgotten crop, is on the rise again. I looked up what I wrote about it many years ago: "All the evidence indicates that spelt is inferior in every way to our other common grains." I'll stand by that as to its usefulness in pasture farming, even though the farmers who still grow it in eastern Ohio swear by its hardiness and usefulness as both grain and pasture.

But in two important respects, I was and am wrong about spelt and may be wrong about triticale in the long run. People who are allergic to wheat flour can eat bread made from spelt, and so a small but lively market has developed for it. (Coincidentally, Bob Evans's wife, Jewell, and their son Stan started a business that, among other grain products, brought out a tasty spelt bread in the nineties.) Two of my farmer friends, Steve Gamby and Ralph Rice, are growing spelt because there is a demand for it from organic markets and from Europe. Why? Get ready for a good laugh. The genetic engineers of our more common grains haven't had any reason to monkey around with spelt. There's not enough market for the big boys to bother. Both Gamby and Rice also think it makes good feed. Mr. Gamby is even using it in place of soybean meal in his dairy cow ration.

Emmer is another ancient, wheatlike grain that is now rare, if grown at all. But I have learned my lesson. I am not going to say it is inferior in every respect to other grains. Tomorrow the consumer world might decide that emmer is the most nutritional grain not available and so demand that some of us crazier farmers try to grow it. Sounds like a good idea to me.

18

Other Plants with Pasture Potential

Besides legumes, there are other broad-leaved plants (forbs) being scrutinized for pasture farming. Some of them we know more commonly as garden vegetables—turnips, rutabagas (swedes), kale, mangels, potatoes, pumpkins, squash, and melons. In fact, while I am not yet convinced of the practicality of most of these plants on larger, commercial pasture farms, on tiny garden farms like the one described in the fourth chapter, using the garden surplus as part of the pasture rotation is eminently practical and economical. Your family cow, or fattening beef calf, or pigs, or sheep, or riding horse can clean up surplus vegetables or eat vegetables specifically planted for them in the garden paddock.

Turnips, Rutabagas, Parsnips, and Mangels

The practice of feeding these "roots," as they were commonly called when used as livestock feed, was traditional in northern Europe, and immigrant farmers brought them along to America with their cultural roots. Some feeding of roots still goes on commercially in Canada or wherever the weather is too cold for growing corn profitably. Where corn has become the prevailing crop, it generally takes the place of roots. Corn provides more nutritional food for animals at less cost. But that "less cost" conclusion is based on harvesting the roots, hauling them to the barn, and feeding them out of storage, not pasturing them, which is decidedly cheaper.

But if fed fresh in the field, instead of held in storage for three weeks, these roots, particularly mangels, can cause diarrhea in livestock. Corn silage can cause diarrhea no matter how long you store it. Some of the loose bowel effect can be avoided by not pasturing roots until they mature—that is, until winter. But the northern farmers in Europe or Canada did not think much of sending their cows out to graze in the bitter cold. They wanted their animals in a barn because they wanted the comparative comfort of the barn themselves. They didn't have snowmobiles and insulated boots in those days.

Modern graziers, particularly beef and sheep producers, look at the problem in a different way. The labor of digging up roots, hauling them to the barn, and chopping them up for animal feed appears to them as too costly and labor intensive. Even animals kept in the barn at night can be turned out on milder winter days to do the root harvesting themselves. The old worries about the animals eating too much dirt with the roots, or of choking on a big old mangel, are considered overwrought. But the problem with grazing turnips in winter is that the animals can't usually get the roots out of the ground when the soil is frozen.

Mangels (and sugar beets which can also be fed) have the advantage of not tainting the taste of the milk like the other root crops. Mangels also keep better. But smaller turnips and rutabagas are easier for animals to eat. Usually the roots are strip-grazed in front of electric fence moved every day or two. Turnips and rutabagas stick out of the ground enough for the animals to pull them up or bite them off. Usually the leafy tops are eaten first, and later the roots. Bob Evans says it took several days for his cows to get used to the idea, but then they grazed turnips readily.

The advantage of roots is that they make highly digestible food in high summer, when cool-season grasses are at their worst, and again in winter, when they have the potential to provide pasture until spring. The disadvantage is that they are very watery and low in fiber and should be fed along with grass or hay. This is not an argument against growing roots, says Bob Evans. He has had cattle grazing turnips sixty days after planting and thinks feeding roots is a good, practical cost cutter.

Parsnips are probably the best root for livestock but are rarely used because they don't stick out of the ground as much as turnips and rutabagas. Historically, dairy farmers in Europe fed them, especially on the islands of Jersey and Guernsey, but it appears that the parsnips were harvested by human hands and stored in the barn. Parsnips' advantage, as gardeners know, is that they will survive in the soil over winter, and in fact taste better in February than they do in December.

Early-maturing varieties of turnips are grown for summer grazing and late-maturing ones for winter grazing. Rutabagas mature slowly and are sown in late spring for fall and winter grazing. They sprout and grow about as well when broadcast on top of the ground as when planted in it if moisture is plentiful. Both have found more favor with beef and sheep graziers than with dairy graziers because these roots can give milk an off-flavor. Dairy cows should not be fed them for the two hours before milking. It is better to graze roots after morning milking and then put the cows on hay or grass in the afternoon. Sheep appear to like rutabagas or swedes better than turnips, according to Morrison's *Feeds and Feeding*. I like rutabagas better myself, so why wouldn't sheep?

My hesitancy with root crops in commercial-scale grazing is the same as with warm-season grasses. Clover and alfalfa are more nutritious, palatable, and permanent, and properly managed they are better pasture for both late summer, fall, and early winter. For late winter pasture, I think corn has a greater potential than roots.

Garden farmers know all about turnips and rutabagas and parsnips for human consumption. For grazing they need to think of them only in larger quantities than they need for the table. If, for example, a garden farmer is raising the family's meat, dairy products, and eggs on a five-acre field divided into ten half-acre paddocks, one of the paddocks would be the garden paddock. A half acre is more vegetables than a family usually needs, but planting that much requires little increase in labor. The animals do the harvesting of the surplus. Early gardens can also be double-cropped to vegetables for winter grazing. For instance, turnips and rutabagas can be planted after early peas, beans, and potatoes, or where a strawberry patch is being taken out of production after June harvest.

Kale, Rape, Cabbage, and Kohlrabi

Kale and rape can survive freezing temperatures quite well, at least for a while, and so have been commonly used for winter grazing. The practice is in a sort of renaissance because of the interest in year-round grazing. Rape is best for sheep and hogs, say the books. It used to be sown into knee-high corn for winter pasture and may again be used that way in the future. Rape works well with commercial pasture farming but probably should not be considered for garden farms since it is not generally eaten by humans. Why grow something not appealing to either humans and animals? Besides, kale is just as good for grazing and is a garden favorite, especially in the mid-South, cooked like cabbage. The first time a plant of kale grew in our northern Ohio garden, it was from a seed that must have accidentally gotten into my mother's cabbage seed; she did not know what it was. When it persisted vigorously green well into winter, she joked that it must be something possessed of an evil spirit. Not until my wife, a Kentucky girl, introduced me to kale did I know how good it was. I have seen one of her brothers eat nearly a half a bushel of the stuff at one meal. Of course, after cooking down, a half bushel becomes merely a potful, but still . . .

The fact that kale will persist, even after suffering temperatures down to twenty degrees Fahrenheit, makes it excellent for winter grazing. Kale also sprouts and grows well from surface seeding in early fall. We usually rely on seeds falling off last year's plants in midsummer to supply our fall-winter garden crop. Plants will sometimes survive winter, bloom again, and make seed the second year. Take milk goats or cows off kale or rape at least two hours before milking to avoid off-taste in the milk.

Kohlrabi, on the other hand, does not seem to give milk an off-taste. Because the bulblike stems grow above ground, they are easier for animals to graze, too. Kohlrabi is good for sheep and dairy goats.

Cabbage is good livestock feed. Like various roots, cabbage used to be fed as a tonic to give show animals, especially sheep, that special glow of good health that attracted the judges' eyes. Young people getting their animals ready for the fair might want to try that again.

Pumpkins, Melons, and Squashes

Pumpkins, melons, and squashes are relished by livestock and in the past were grown in cornfields for winter pasture along with the corn fodder. I grow some, along with squashes and melons, on the edges of the cornfield, where they get enough sunlight to grow okay. After corn harvest, when I turn in the animals to clean up the fodder and ears I left, the vegetables disappear, too. Some people believe it is better to break the pumpkins so the livestock can eat them more easily. My livestock don't need help. They bite into pumpkins no matter how big, about like we bite into apples. In folklore, pumpkin seeds are thought to be good wormers.

In large gardens there is almost always a surplus of vegetables. Instead of having to clean them up, turn in your animals. If you are addicted to growing far more tomatoes than you can use, as we are, this is a way to get rid of them. Cattle will eat them but not as eagerly as chickens will. Goats will eat them because goats will eat anything, even electric insulators. Just kidding.

Potatoes and Sweet Potatoes

Cull potatoes or surplus stored potatoes that have deteriorated too much for sale have long been fed to fattening beef. They are about as good for this as corn silage. Hogs will do well on potatoes if they have some other feed for protein. But for hogs, unlike other livestock, the potatoes need to be cooked. Corn is so much more nutritional. Since in either case the crop has to be cultivated and planted annually, corn is preferred. In northern Europe, where corn doesn't grow, special potatoes for livestock are still grown. For the garden farmer who grows potatoes for the table, feeding the culls or surplus to livestock makes good sense.

Hogs will root up sweet potatoes and eat them. Back when people were careful about their money, a specialized kind of grazing developed where sweet potatoes were grown commercially in large fields. If there was a surplus that could not be sold, hogs were turned into the field to clean them up. Since in harvesting any kind of potato the field has to be dug up, letting hogs root up the field was not a problem. They became the primary tillers for the next year's crops.

Sweet potatoes are much higher in dry matter than other root crops, but it still takes about four bushels to equal the nutritional value of one bushel of corn. They can be fed to other farm animals as well, including horses, as part of their feed ration.

Puna Chicory

More-or-less exotic new introductions are being offered as pasture plants, like Puna chicory, which was developed in New Zealand. Regular chicory that grows wild along roadsides is considered a noxious weed in some states, but Puna (I violate my rule not to use varietal names because Puna was developed specifically for pastures and is specifically exempted from regulations that prohibit the planting of common chicory) is relished by domestic and wild animals and does not seem to present a weed threat. The downside to Puna is that frost flattens it, so it can't be pastured after killing frost. I suppose that new varieties will come along that will be hardier.

I personally don't think common chicory is much of a weed threat. Its blue blossoms flanking the roadsides are pleasant to look at. The roots can be used as a coffee substitute, and in some parts of Europe it is grown commercially. I suppose Puna roots could be used the same way.

Chard

If you grow some of the new colorful chards for your table, you know how easy it is to plant more than you can eat. Chard is an excellent graze for chickens. That's the only reason my mother grew it. She could never get any of us to eat the stuff. My daily job was to pull up an armload of chard and toss it into the chicken yard. It never dawned on any of us to plant some where the chickens could graze it themselves. I am sure chard would also be relished by livestock, because in our garden today it is the first green that deer and rabbits ravish. Even I like the new colored varieties. Chard is a tremendous producer, and a patch of it can provide on-and-off grazing through the growing season. Whack it off and it grows back. It is also easy to plant. Just throw the seed in the general direction of the soil, and when rain comes, get out of the way.

19

Weeds—the Good, the Bad, and the Beautiful

I am a weed watcher in the way others are birdwatchers. Being a weed watcher is not something a farmer ought to admit to. Farmers aren't supposed to watch weeds; they are supposed to kill weeds. The weed I am watching at the moment, however, refuses to die. The dandelion. Every spring in my hayfields, dandelions run riot. There is a field nearby that beats any stand I have ever not tried to grow. You can't imagine anything so beautiful as this field in May with its dense stand of golden-blooming dandelions. No doubt there are ten thousand gallons of good wine going to waste there, and five thousand jars of dandelion honey, and maybe a hundred tons of what is thought to be the second most nutritious salad green in the world. The farmer involved with that field has perfected the cleverest method of double-cropping (actually triple-cropping) that I've seen, and he doesn't even try to do it. It just happens. He pretends to no-till wheat and soybeans in rotation. After planting beans down through the golden dandelions in May, he sprays, which burns back the weed enough so that he can pretend to take a crop of beans off at harvest time. Then the deeply rooted dandelions come back, drawing minerals up from deep in the soil, forming an erosion-proof

209

covering, down through which he no-tills wheat in the fall. If he had a fence around this field, he could graze dandelions in spring and fall and not hurt his beans or wheat.

I am not making up a bit of this. Here's a case of sheer genius, even if not yet recognized, like all genius when it first appears. And what the farmer has done so far is only the beginning of possibilities. I have been talking to one of the world's authorities on the dandelion, Peter Gail. I wanted to find out why my sheep eat not only dandelion leaves and blossoms but even the old white stems with fuzzy seed heads sticking above the alfalfa. He wasn't sure, either, but that's how I found out that the dandelion is considered by some botanists as the second most nutritious salad green in the world. That of course means sheep are smarter than most humans. The first most nutritious weed, get this, is lambsquarters, and the third is redroot pigweed (amaranth), two of the worst weeds for grain farmers. Read it and weep. Mr. Gail, who is not without a nice sense of humor, guessed that since livestock like dandelions, they would like the old white stems too because they are even more bitter than the leaves.

In early spring the leaves are not so bitter, and many people love them as much as livestock do. Mr. Gail said it was smart to graze dandelions but that more money could be made by harvesting them in early spring for sale to grocery stores that deal in herbs. Then graze what is left. Meanwhile, you could make several barrels of good wine per acre from the golden blossoms.

A firm in Maine has been selling canned dandelions since 1886. Americans spend three million dollars a year buying dandelions to eat. While the lawn lovers try to kill the things with poisons, other people buy the seed from nurseries and plant it to grow their own supply. Mr. Gail thinks that this is very funny, but it is enough to make a farmer or a suburbanite cry.

Dandelions are an effective diuretic. A common name for the weed in England is "piss-a-bed."

All of us who call ourselves farmers need to take stock here. This is not just passing humor but a golden (get it) opportunity. How many millions of dollars do you think are spent on diuretic pills each year? The alternative is staring us in the eye. Finally, a crop that would make money. And the cost of growing this wonder herb is exactly zero. Just pretend that you are no-tilling wheat and beans.

There is irony in my effort to come to peace with weeds. I have learned the hard way that cultivation is responsible for most weed problems, but I won't quite (yet) quit cultivating. I know there are exceptions (like mesquite in the Southwest), but on my farm if I committed all my grazing land to permanent grass and clover, all the weeds that grow here so far would be relatively easy to control. Cultivate for weed control and you make sure that you will always have weeds to control.

The first time I should have learned that lesson was when a botanist informed me that poison ivy does not grow where soils have never been disturbed. I was charmed (although not convinced). If true, does nature somehow "know" that cultivating the soil is a great threat to earth and so she unleashes poison ivy on the cultivators? Lovely.

At any rate, pasture farming has taught me that undisturbed grass and clover sods will, with the help of grazing and an occasional mowing, banish most weeds and keep the others under adequate control. Weeds are nature's way of starting the long road back from cultivated land to permanent woodland or to permanent prairie where rainfall is not sufficient to grow trees. André Voisin, often referred to in this book, said that after being cultivated, soil needs as much as a hundred years to regain its natural state. The harder we try to kill weeds, the more determined they are to live, even finally becoming immune to weed killers. Watch farm fields that have been beaten down with cultivation and weed poisons for years. The weeds do not go away. Let there be any let up in the beating and the weeds spring up worse than ever.

I learned the truth of all this by fighting Canada thistles, which are not Canadian or American but an immigrant from Europe. They are the most pernicious weed I have to contend with. I grew to hate them as a child when one of my jobs was to hoe the alien invader in our pastures. Decapitating Canada thistles is ludicrous because the network of roots just sends up more shoots, but my parents believed firmly that idleness was the devil's workshop and idle boys the devil's favorite workshop. I thought Canada thistle patches were the devil's workshops. Our Canada thistles thrived only in patches, not everywhere. My idea of showing that I was tougher than my sister, Marilyn, was to run barefoot through them. It only slowly dawned on me that the patches marked the places in the permanent pasture where hogs had rooted up the sod. That pasture is still pasture, as it has been for over a hundred years, and it is still the domain of my sister Marilyn, although she isn't quite that old. Without cultivation or rooting hogs and with yearly grazing and mowing, there are no Canada thistles growing there today.

But I did not put two and two together until I started to experiment with grass farming. One of the first things I learned was that in a managed rotation system, sheep would eat Canada thistles when they were young, if nothing else was handy. So would several bugs. And some kind of disease attacked the thistles when they tried to grow in a vigorous grass sod where they were weakened by regular grazing and mowing. The thistle leaves turned white and later dwindled away. Every year a few new ones start up someplace else but soon disappear under the onslaught of hungry sheep.

But in my temporary pastures where some cultivation is done, the thistles thrive even though occasionally pastured and regularly cut for hay. Spraying weed killers is not practical in this situation because that would kill the clover too. Only by cultivating between the rows of corn and spot spraying when a paddock is planted to corn every fifth year can I hold the devils at bay. And then they get busy in their workshop pastures the next year. I am well aware of the irony in the situation. The problem is not the thistles but my insistence on disturbing the soil to grow a little corn every year.

The main general rule in controlling weeds is to note whether they are annual—that is, they come up from seed and die in the same year; biennial, in which case they grow one tap root and, after their second or seeding year, dwindle and die; or perennial, growing and spreading mostly by root. Bull thistles *(Cirsium vulgare)* are biennials. They germinate and start growth one year and flower and die the next. If you chop one out with a hoe or spade at least two inches down on the root, it will usually not grow back. So, like annuals, they are usually easier to control that perennials. Just don't let them go to seed. But, you can cut all the way to China on a Canada thistle root and it will merely send up more plants. Marilyn and I made our spending money as children chopping out bull thistles and wild carrots (what urban people refer to so eloquently as Queen Anne's lace). Dad paid us a nickel a hundred and trusted our counting. He knew he could trust us because each of us watched over the other's addition like a teacher in math class.

Perennials that spread by way of roots are difficult enough. But perennials that also go to seed heavily are the worst. In August I can lie on my back in the pasture and watch the lovely seeds of Canada thistle go floating over, each on its own silken parachute, drifting in from neighboring annually cultivated wheat fields where thistle control is particularly difficult. Like the poor, Canada thistles will be with us always.

The following is a list of the not-so-bad weeds on my farm and another list of the worst ones, from the viewpoint of pasture farming. There are many other bad and not-so-bad weeds, of course, but that would require a tome heavy enough to blot out a Canada thistle patch. For practicality, I'll ignore many of those that are not a pasture problem because the livestock eat them readily. Tree seedlings, for instance, can be the worst weed of all as nature fights to return pastures to woodland, but animals will eagerly eat nearly all of them. Make your own list for your own pastures and study the weeds for weaknesses that you can use to control them or for benefits that you can take advantage of.

Some Good Weeds—Well, Sort of Good, Anyway

Dandelions may be the curse of garden and lawn, but they are for pasture farming a beautiful and nutritious weed. Some authorities say that they are nature's richest source of beta carotene, which the body converts to vitamin A. Animals eat them readily, including the blossoms, as I have pointed out. Dandelions make wine revered by many people. I sometimes wish I liked dandelion wine because I could make eleven billion gallons of it in our county alone.

Because dandelions are so prolific, they are one of the few weeds liked by livestock that will endure in the pasture. Animals graze other weeds that they love, like lambsquarters, as fast as the weeds can grow. But not dandelions. They tend to get out of hand in April and May unless grazed very hard. In paddocks reserved for hay and pastured only infrequently, dandelions can run amuck and make the field look, well, seedy, when all those fuzzy seed heads stick out above the even green surface of the hay. I used to worry that too many dandelions would overwhelm the hay. But I have learned to relax. The dandelions only dominate in spring when the other pasture plants are growing vigorously, too. Being a biennial, the individual plants dwindle in

June of their second year and are no longer a worrisome factor. The new seedlings in the fall can be pastured. The long tap root makes the plant fairly drought-proof. There are times in dry weather when the best pasture available is dandelions. Those deep roots also draw minerals up from the subsoil. When the plant dies, the root rots away, leaving a hole for rain to move down into the soil rather than run off too fast.

Narrow-ribbed plantain is considered to be the most palatable plant of all for livestock in traditional English pasture farming. Extensive experience—even scientific testing—over centuries seems to have proved this belief. Plantain is even considered more palatable, if you can believe this, than white clover. English graziers actually plant plantain in their pastures along with some of the other weeds listed below, as Englishman Newman Turner tells in his valuable book, *Fertility Pastures and Cover Crops,* first published in 1955. But in our current suburban culture, plantain is just an obnoxious lawn weed. Think of the irony. Millions of dollars are spent each year on noxious poisons to kill dandelions and plantain, two of the most beneficial plants for turning grass into flesh. Not only that, but the two make a great combination, plantain coming on in June just as dandelions are dwindling.

Anyone familiar with the ways of lawns knows how to grow dandelions and narrow-leaved plantain. Simply establish a bluegrass/white clover lawn and they will come. When that happens, pretend that you want to kill them. Mow like hell. The more you mow, the faster they will grow. If you have to stop long enough to fill your mower with gas, you might as well start over again. I once marked a plantain and studiously cut the flower stem off when it grew up. Overnight (and I mean overnight, literally) a second stem sprang up. I subsequently cut the stem off four times before the plant gave up.

If you can trick plantain into thinking that you hate it by clipping the pasture once or twice, the weed will make good pasture through the hottest, driest part of midsummer when the bluegrass dwindles. Incidentally, if you are willing to put up with occasional visits from men in white coats, Newman says to sow narrow-leaved plantain at one pound per acre, but he prefers, for dairy cows, two to three pounds per acre. I don't know where seed would be available today. Perhaps from terrorists?

Simply by pretending to want to kill it, I now have good stands of narrow-leaved plantain in my hay paddocks, which occasionally I use for pasture, too. It took me a couple of years to turn my brain around, but when I see a nice stand of plantain growing in the clover, I am as pleased as when I find a nice stand of timothy. The seeds find their way into the permanent pastures, too, to make a nice stand, but the sheep gobble it up there, and I must wait until another batch of seed moves in from the hay paddocks.

Broadleaved plantain, another yard and garden pest, won't last long in pastures, either, because sheep love it. I had a great patch of it volunteer one year near a haystack. Because the leaves are rather broad, and because the plants tend to grow close together, I feared the weed would smother the bluegrass and clover. But the animals decimated it quickly, and only enough grows every year here and there in the pastures now to give them a taste.

It is interesting to see the sophistication of Turner's seeding recommendations for temporary leys in England. Remember, this was half a century ago. For midsummer pasture, for example, he recommended:

Lucerne (alfalfa) 6 lbs. per acre
Chicory 6 lbs.
Timothy (two varieties) 3 lbs. each
White clover 3 lbs.
Burnet *(Poterium sanguisorba)* 3 lbs.
Late-flowering red clover 2 lbs.
Meadow fescue 4 lbs.
Perennial ryegrass (two kinds) 6 lbs. each
American sweet clover 2 lbs.
Sheep's parsley *(Petroselinum sativum)* 2 lbs.
Caraway 1 lb.
Narrow-leaved plantain 1 lb.
Broad-leaved plantain 1 lb.

That's 'way more total seed than American pasture farmers would use today even if they could get all those species. Too much alfalfa, ryegrass, white clover, and timothy, in my opinion, but maybe back then seed was not available with high germination rates. I'd drop out the sweet clover. Even Turner admitted it is one of the least palatable plants for cows. He included it because he evidently thought that it was good to plow under to make soil more fertile when in fact it is no better than other deep-rooted legumes for that purpose. He considered caraway, which is known to help dyspepsia, to be an effective antibloat plant. In his experimentation, his cows loved parsley almost as much as plantain. Because of the apiol in it, he believed it to have a "potent beneficial effect on kidney and bladder complaints" (which is why parsley became a common garnish with human meals) and to be "also specific in female reproduction disorders." In fact, he recommended a higher rate of parsley in the pasture for cows having breeding problems. He planted burnet because its palatability was considered on a par with plantain and because of its ability in English acid soils to root deeply into the chalk subsoil to bring calcium up to the surface, thus acting as a natural lime spreader. Obviously, English graziers of his time were more knowledgeable about weeds than most of us Americans are today.

Lambsquarters and *pigweed* thrive in cultivated soil, as gardeners and farmers know, but they are scarce in pastures. The animals will eat every chance plant that grows. When sheep are turned into cornfields in late summer, they will eat all the lambsquarters and pigweed that have grown up among the cornstalks. Once people from India were visiting the Rodale Experimental Farm at the same time I was. Suddenly they began chattering excitedly and pointed to a weed in the fencerow. They asked if it was okay to pick it. Lambsquarters. They explained that it was a great delicacy in their country.

Lambsquarters is often at the top of the list of the most nutritious plants for humans and animals alike. When it grows in your pastures, rejoice. If you find plants going to seed in cultivated fields, pick off the seed heads and scatter them over your pasture. No grain farmer will ever object to your stealing his lambsquarters. I have noticed that lambsquarters will grow profusely where a haystack stood a year earlier.

Redroot pigweed, or, more correctly, amaranth, shares "most nutritious" status with lambsquarters and dandelions. So-called primitive Americans gathered amaranth seed for eating, and apparently it was cultivated even before maize. The Rodale Experimental Farm once conducted an extensive study of amaranth as a farm and garden crop. That research produced good information, still available from the Rodale archives. But there is irony in it for me. The amaranth trials focused on the seed as a harvested, stored grain, in the same way that we have focused on wheat, oats, and other cereal grains, rather than as pasture plants that can be harvested by livestock (and a little for the grazier) for leaves, roots, and seed. If you have a field "infested" with pigweed, all you have to do is turn farm animals into it, and they will turn the problem into meat and dairy products. Livestock will eat the leaves and seeds, and hogs will eat the roots.

Chickweed in the garden is a worse weed than dandelions, but, like dandelions, it makes a salad decent enough that some commercial demand exists for it. A salve is also made from chickweed that is claimed to be good for all skin disorders (from S.A.E.G., 1362 Hwy. 129, Canaan, IN 47224). Chickweed is a relatively new weed in our vicinity. I made its acquaintance while driving in northern Ohio in December. I passed a field that was almost solid green. The sight was dramatic enough for that time of year that I stopped to take a closer look.

Later, when chickweed found our garden, I learned it has awesome survival powers in cultivated soil. Rotary tillers only make it grow back thicker. A gardening friend got so overwrought that he bought a flamethrower and "toasted the damn stuff to death."

When I started thinking about what plants might make winter grazing in the North, I naturally thought of chickweed. I had no trouble grubbing up a bushel basket full from the garden in December. I fed it to the sheep. The first day they would not touch it. The third day they ate it all. So now, against my better judgment, I have a start of chickweed in one of the temporary paddocks. I'm not worried about it getting out of control because it won't last in a grass sod. But that's also the reason it may not be practical for pasture.

Milkweeds are very bitter to the taste and are even considered poisonous by some authorities, but livestock will occasionally bite off the tops. There must be a reason for that, but I hesitate to turn to the claims of herbal remedies to offer one. One books says in bold letters: Do Not Eat Milkweed. Other herbalists list it as a substitute for asparagus, but only if you cook it with at least one change of water. I have been through that "one change of water" maneuver with other wild plants, and I will just say that you would be wise to stick with asparagus.

But I like having milkweeds in the pastures because with grazing and an occasional mowing, they don't become a pest as they do in cultivated fields. Also, the flowers are beautiful. And the plant, with its sticky milky ooze, offers commercial possibilities as a substitute for rubber or as a biofuel. The seedpods, with their silken seed parachutes, can be used in fabrics. The milky stuff evidently has potential for medicine. I once applied it to a wart and, much to my surprise, the wart sort of deteriorated and went away. During World War II, farm children earned a little money harvesting the pods, from which the seed silks were extracted for use as filler in life jackets. Also, the milkweeds draw several beautiful insects, especially the monarch butterfly. I have tried to introduce the eye-catching orange-blossomed species, butterfly weed, but it doesn't like my pasture environment. It grows better in dry, almost barren soil. Where it occurs naturally, butterfly weed is one of many plants that can turn a pasture into a flower garden while turning grass into flesh.

Yarrow volunteers naturally in pastures but never seems to overpopulate. It is quite bitter to the taste; livestock eat it, but only sparingly. Herbals claim medicinal value for it, especially as a wound-healer, but the herbalists invariably retreat into the subjunctive mood when citing curative powers for wild plants like yarrow, so I remain skeptical. I presume yarrow is healthful for livestock in some way or they wouldn't eat the bitter stuff.

Japanese honeysuckle does not commonly endure in colder northern regions, at least not yet, but there were times when I was living slightly farther south that I would have named this vine the worst weed of all. It can vine up trees and kill them. It can turn a fencerow into a jungle. But I didn't realize that the fencerows and woodlots where it ran rampant were places grazing animals could not get to. I didn't figure out the significance of that at all until I became acquainted with Bob Evans. Bob, looking for plants that could be grazed in winter no matter how deep the snow, considered Japanese honeysuckle. He knew cows would eat it. He knew that it stayed semigreen through the winter in the Ohio River valley. He pestered agronomists until they ran some analysis on the nutritional value of the weed. Guess what. "Even the dead brown leaves have a crude protein content of over 23 percent," he told me.

Many other weeds of potential usefulness deserve mention. *Sheep sorrel*, a problem in gardens, has a spicy, lemony taste and makes a salad garnish some people enjoy. Grazing animals enjoy it, too, and I am always glad to see a stand invade the pasture. It doesn't last long. Animals will graze nutgrass, an abomination in low areas of cultivated fields. The corms on the roots are tasty, and hogs will search them out.

It is obvious to the garden farmer that many weeds in a pasture can double as garden vegetables or healthful herbs. If grass farming develops into a universal food system the way gardens have, there will be much more experimentation in this direction.

Bad Weeds

Thistles. I have already discussed the Canada thistle and the bull thistle. The latter is almost indistinguishable from the pasture thistle, *Cirsium pumulum*. Both, but especially the latter, have attractive blossoms and on a very small pasture farm you might want to allow a few to bloom just for their looks. Another reason to allow a few to grow in spring is that song sparrows and similar sparrows like to nest at the base of these thistles. The music of song sparrows is well worth risking a (small) invasion of bull thistles. Just be sure to cut the thistles after they bloom and before the seeds scatter on their fuzzy parachutes to all parts of the farm. Cut bull thistles and pasture thistles two inches below the ground surface. If you only mow them, they will develop secondary stems and blossoms and keep doing so after every mowing until the last blossoms barely stick above the soil surface. Livestock won't touch them. The thorns are too long and stiff. Once a paddock is rid of them, it is not difficult to go over it with a hoe every year and chop out the few trying to sneak in.

There are also so-called sow thistles, which are a pest in gardens because they spread by both root and seed like Canada thistles. But livestock will graze them fairly well. The thorns are small and not so prickly. They are not really thistles and resemble dandelions somewhat.

The musk or nodding thistle is very bad but fortunately none have yet drifted onto our farm. Walking with my cousin on his farm once, he spied a nodding thistle in the fence line. He swore. He called it a "Lonsway" thistle after a neighbor of many years ago whom he considered a little too liberal on the subject of weed control. He dug the Lonsway out with his pocketknife, carried it to the road, cut it into little pieces and ground the pieces into the pavement under his heel. I've read that mules will eat nodding thistle but wonder if the animals have to be half-starved to do so. The spines are not as nasty as the ones on the bull thistle but nasty enough.

Sourdock (or curly dock) grows a seed head that must contain a trillion seeds or so. If just one plant is allowed to go to seed in a ten-acre field, the next year three acres will be covered, and the year after that the whole ten acres. I exaggerate only a little. *Burdock* has large burrs that get caught in sheep's wool and practically ruin it. The weakness of both docks is that they are only biennials. If you cut them out two inches below the soil surface, they are done for.

Faced with an old pasture occupied by an army of sourdock, the best control method I have followed is to let them go to seed. Then mow and rake into windrows before the seeds shatter off, and burn them. Repeat when the plants regrow. The next year, pasture early because sheep and cows will eat the early growth. You may have to do the mowing and raking bit again. In a year or two, you will reduce the army to a mere battalion and then you can start chopping out the remainder by hand-hoeing. Eventually, only a few will grow every year, but it will be of utmost importance to walk the pastures and pull or cut them out before they go to seed. Or, if in seed, cut them and carry them out of the field and burn them. You can't just let them lie on the ground, because if the tap root can get any kind of purchase in the dirt, or in wet grass for that matter, it will continue to grow and the seeds might mature. I don't know how many thousand I pulled out by hand when I first acquired our farm, but now there are only a few to contend with every year.

With grandchildren and partner families entering the picture, we can now field a veritable army of hoe wielders to fight the veritable army of sourdock on new land we have acquired. "Dock Cutting Day" in May has become something of a tradition in the family.

Burdock, with its huge rhubarblike leaves, is not quite as persistent as sourdock. The plants have to be hoed out because pulling is difficult. There is a commercial market for burdock roots, but we never have enough roots or time to make that worthwhile. Sheep will eat burdock, but not eagerly. If you are looking to sell wool at a good price, you must keep burdock out of the pasture in any case.

Queen Anne's lace, or *wild carrot,* may seem charming to wildflower fanciers, but it is a pest in pastures. And if dairy cows eat enough of it, their milk will have an off-flavor. In the rural tradition I was brought up in, a farmer needed to keep sheep if for no other reason than to control wild carrots. The truth of the tradition was proven by my father, who did not like sheep (or tradition) and sold his flock a few weeks after Grandfather Rall, who still ran the show, died. Shortly thereafter, wild carrots all but took over the pasture.

We were taught that wild carrot roots were poisonous, but I can find no support for that anywhere. Actually, the roots taste about like garden carrots, only a bit too strong for me. The roots are white rather than the orange of garden carrots. The two plants are otherwise nearly identical. I think we were told that wild carrot roots were poisonous because the ferny leaves of the plant look quite a bit like poison hemlock when the latter is in its early stages of growth. There is no future in eating hemlock, as Socrates understood.

Mullein sometimes proliferates in pastures, especially overgrazed pastures. Animals won't eat the soft, furry leaves. But those leaves make an emergency toilet paper if you get caught out in the pasture and have a need for some. The weed is also attractive in bloom with its tall spikes of yellow flowers. If you don't cut them out they can multiply too much in permanent pastures. But they won't compete with a vigorous pasture sod the way sourdock will. Herbalists insist that mullein has medicinal value.

Ironweed likes low, rich, moist soil. The plants bloom quite beautifully and a few sticking up in the pasture showing off their purple blossoms have a flower garden charm that offsets their potential invasiveness. In my wooded pasture across the creek (Paddock E) I once amused myself by mowing all the brush and weeds in one section except the purple ironweeds and the yellow goldenrod. The result was a very unusual and attractive Garden of Eden. Ironweed is fairly easy to control with an August mowing. The sheep and cows will eat a little. Tim Kline says his meat goats eat it readily.

Ragweeds are no doubt more of a pest for asthmatic humans than for grazing livestock. Cows and sheep will eat common ragweed fairly well when it is young. They will absolutely devour giant ragweed at all stages. The latter has become one of the worst weeds for crop farmers after the cessation of grazing and haying on so many farms today. I've seen stands of it grow so tall and stout from all the fertilizer grain farmers use that the plants stopped giant grain harvesters dead in their tracks. Since in my experience animals like giant ragweed leaves better than they do corn leaves, is there something here we ought to be studying?

I have discussed *bindweed* (sometimes called wild morning glory). Like Canada thistle, only more so, it loves cultivated fields, and the more you cultivate it, the faster is spreads. The blossoms, which vary in color from purple to white to deep maroon, are beautiful to everyone except farmers and gardeners. A vine climbing up cornstalks can actually overwhelm the corn plant. But a mass of vines in blossom clinging to tall cornstalks is downright pretty even if they can make harvesting a living hell by clogging up gathering chains on older combines. In the "old days" when I was cultivating weeds in corn with a tractor cultivator, so much bindweed would get caught on the cultivator shovels that I would have to stop every round to pull them off. The earliest advertisements for herbicides inevitably showed pictures of angry farmers trying to pull bindweed vines off their equipment. The herbicides changed that for a while, but now the pest is showing signs of immunity to the chemicals.

In a pasture farming world, bindweed is not a bitch but a queen, since grazing animals eat it readily. I used to hate the stuff in hayfields until I found out the cattle and sheep would eat it right along with the legume hay. Nor will it survive in heavy sod pastures. Also, it is very susceptible to frost. I've seen it start to gain a foothold in a sod, and then late frost set it back so much that it was completely smothered out by the grass.

Garlic mustard, another invasive plant from Europe, is savagely replacing native wildflowers in eastern American woodlands. Naturalists are having fits because trying to kill it would involve killing the wildflowers. It grows rank, up to three feet tall, with puny little white blooms. It all but overwhelmed our woodlot until I turned the sheep in. They ate every bit of it. Of course, they also ate the wildflowers. But the wildflowers will come back. They have survived sheep before.

Poisonous Plants

The subject of poisonous plants is a never-never land that I address only hesitantly. There are at least twenty plants with toxic capabilities that may show up on farms. Listing them will probably only increase the fears of paranoid modern society to little good effect. I know for a fact that grazing animals will eat some plants that are listed as deadly poisonous and not be harmed unless they eat too much. Oak leaves and acorns are supposed to be poisonous. My sheep love acorns, as do hogs, deer, and squirrels. Apparently, cattle get poisoned (the tannin in the oak leaves and acorns does it), but only in special situations. Young cattle not getting enough to satisfy their growing appetites have died from feasting on acorns. *Feasting* is the key word. Older cows, eating lush green spring pasture, crave fiber. If they can get to oak leaf buds they will not eat just some, they will feast. But those situations are rare.

I have at least fifty oak trees that my animals can get to, but there are no longer branches low enough for them to reach. They ate all the low branches off! I regularly sample acorns, looking for trees, especially white oaks, that produce fruit that is less bitter. Western Indians practically lived on acorn flour and bread, soaking the tannin out first. The chinquapin oak's acorns have little tannin and are quite mild to the taste. Red maple leaves are supposed to be poisonous to livestock, too, but again my animals can't reach our red maple branches anymore because they ate all the lower branches off.

Some other dangerous poison plants in pastures:

Buttercups (Ranunculus sp.) are beautiful to look at but are dangerous in pastures. However, they prefer a wettish soil and usually don't grow in good, well-managed pastures used for rotational grazing.

Larkspur poisons several thousand cattle on western rangelands each year, but sheep are unaffected. Ranchers pregraze larkspur with sheep or keep cattle away from larkspur in its earliest stage of growth, when it is most potent. According to the Agricultural Research Service, scientists, while admitting they don't know why cattle are attracted to larkspur, have noted that rapid eating of larkspur frequently occurs at the onset of stormy weather when barometric pressure and temperature drop, accompanied by high humidity, rain, and leaf wetness. I hope those august scientists don't mind if I find that too much to believe.

I am having quite of bit of unwelcome experience with *poison hemlock,* the latest alien to invade our farm. Poison hemlock is described in weed manuals as one of the most deadly of plants. But my sheep will eat a little of it regularly even in the springtime when they have plenty of good pasture. So what's going on here? I like to think (no proof) that the little bit they eat is killing the internal parasites in them. Apparently, animals have to eat quite a bit at one time to be harmed, at least a half pound's worth, according to several sources. That's a lot of hemlock leaves. The roots and seeds are supposed to be the more deadly parts, so perhaps eating the leaves is not as worrisome as I at first believed. I am almost certain that the two ewes I lost last summer died from overeating it, but by that time, the seeds were forming and were possibly the cause.

Fortunately, hemlock is more or less annual, spreading by seeds, not roots. Timely mowing several times a year will get rid of it. But the weed likes the brushy areas along the creek where I can't mow. I have to resort to a weedeater or herbicides. I am diligently spraying it now that I know what it is, but it has gained a foothold, and extermination is not easy. So much of it is growing in our county that I worry about children. The papers reported that two children in Cincinnati had to be hospitalized after just playing in a stand of hemlock. The plant is so god-awful bitter I doubt any human would swallow enough to die from it unless hell-bent on suicide. (I tasted it.)

Horse nettle, a perennial member of the solanum (nightshade) family, is almost as bad a weed as Canada thistle. In addition, it is poisonous. Equines will eat it, I'm told, but my cows and sheep will not touch the prickly leaves. However, as I have mentioned, they will eat the blossoms. That helps, but not much. Mowing discourages it, and by fall it dwindles away until next year. Mowing and blossom grazing are decreasing its population in our permanent pastures, but it is a slow process. I need a horse again. As long as we had a horse on our pastures, horse nettle was not much of a pest. The next best defense is a good strong sod.

We had a scattering of *cocklebur* on our farm when we first bought it. I pulled every one of them, made piles, and burned them. The young seedlings are listed as poisonous, but sheep eat them. If cocklebur grows in a wheat field, the seeds mix in with the wheat grains and are hard to get out with a seed cleaner. In any event, they spread mightily, but more in cultivated land than in pastures. Undisturbed sod deters them.

Poison ivy is not much of a problem because sheep will eat it readily. Why it will blister humans and not animals is another mystery of life. If you have poison ivy in an orchard, a good way to get rid of it is to pen sheep in.

This is not by far a complete listing. As a grazier, you should try to learn the names of everything that grows in your pastures (it's fun), and then you can check your list against the list of poisonous plants. *Lupines, bouncing Bet, charlock* (a wild mustard but not the succulent mustard, rape, that volunteers in pastures), *white hellebore, white snakeroot, sneezeweed, dogbane, arrow grass, horsetail or scouring rushes, water hemlock, laurel,* and others ought to be mentioned. *Jimson weed* is viciously toxic, but it always grew in our barnyards when I was a kid and the animals never touched it. We were told that the juice was good for poison ivy rash, but I never tried it.

Most of these poison plants do not grow much in pastures. Nor will animals normally eat them unless they are very hungry. You might have to worry if your animals have access to woodland or brushy areas. White snakeroot, for example, is a woodland plant and is so virulent that it can poison a person drinking the milk from a cow that has eaten it. So the books say.

Even *pokeweed* makes the poison lists, although I eat the new shoots in spring just like I do asparagus. In fact, we allow a few plants to grow in our asparagus bed. I've rarely seen poke growing in pastures, because grazing animals eat them in early spring when I do. The seeds in the berries are poison, too, but birds eat the berries, and the seeds go in one end and out the other without harm.

20

Making Hay and Silage

If pasture farming ever reaches its full potential of providing grazing for the whole year, making hay or silage might become obsolete. That time has not come yet everywhere and probably never will completely. For one thing, there will always be a surplus of pasture in early summer, which it will pay to at least clip. If you are going to clip it, you might as well make hay or silage from it for winter feed and drought emergencies. It is, in fact, the advances in hay and silage tools and production methods that aid the advance of year-round grazing and the hope that annual tillage will go the way of the stagecoach.

Making hay has traditionally been labor intensive and the hay extremely vulnerable to damage from adverse weather during harvest. No one foresaw the awesome technology that would transpire in an effort to get hay harvested with no more labor or weather risk than getting grain harvested. The newest mowers can cut any tangle of hay without plugging, crimp it—that is, squeeze out excess juice in the stems about the way an old wringer washing machine squeezed water out of clothes—and windrow the forage in one pass. On a hot day, such forage needs to dry only in little more than fifteen hours with help from the latest tedders. The new balers are equipped with computerized sensors that detect when the hay is dry enough to bale. If the moisture content is over 28 percent, the baler shuts down. Just won't run. The art of knowing when hay has cured enough to be stored just by feeling it or handling a forkful of it, like us old-timers do, is practically obsolete. If the moisture content is not quite high enough to shut down the baler, other sensors activate a sprayer that squirts just the right amount of citric acid on each bale to keep it from molding. (The sensors that work these "miracles" cost about $5,000.) Under the most favorable conditions, these balers can move along in the field at fifteen miles per hour or more! And if the operator of the tractor pulling the baler gets skillful enough with the levers that control the bale thrower, he or she can do a fair job of ranking the bales on the wagon behind the baler without getting off the tractor. In any event, the bales end up in the barn or on trucks in half the time or less as formerly, having hardly been touched by human hands.

These balers will likely be used mostly by commercial hay growers who sell hay because rectangular bales are, or seem to be, easier to transport long distances. On-farm haymaking, on the other hand, is more and more going to big round bales that weigh half a ton each. Haymaking is faster with round balers. The bales wrapped in netting shed water fairly well, and the bales are moved mechanically rather than by hand, avoiding the thankless task of trying to hire temporary workers for haymaking. All farmers thinking of going into commercial grass farming should look seriously at big round bales.

Such hay-making equipment is expensive, of course, and therefore not practical on small pasture farms. In fact, on the smallest "backyard" pasture farms or pasture gardens, a lawn mower or a little sickle bar mower on your tiller, plus hand rakes and forks, are all the equipment needed. And because an operation of this size will be harvesting only very small amounts of hay at a time, the hay can usually be gotten under shelter before rain without needing the speed of these expensive machines.

On a pasture farm of ten to a hundred acres like ours, a small farm tractor, a sickle bar mower, a hay rake, and a pickup truck or hay scoop fitted to the hydraulic front-end loader of the tractor, plus perhaps a baler but not necessarily, are sufficient for harvesting hay. In addition, a rotary mower for clipping grass pastures will usually prove to be more satisfactory (doesn't plug up as often) than a sickle bar mower, although grass clipped by a sickle bar mower seems to grow back a little faster.

With older farm equipment or lawn equipment and hand tools, making hay properly requires at least three days of sunny drying weather in early summer and two in hot late summer. It is not easy to get three or four days in a row of sunny weather in the Midwest, mid-South, and Northeast, especially in May and June, when the heaviest growths of hay need to be harvested. A new weather front rolls through about every three days, often accompanied by rain. The haymaker becomes an astute watcher of the weather map and learns by noting the movements of these fronts on radar to predict the next rain about as well as (and sometimes better than) the National Weather Service can. I used to try to wait for a three-day sunny forecast to cut hay, but such forecasts were rare in early summer and often wrong. Forecasters are very good at predicting the current weather (look out the window), but they are awful at predicting two to three days ahead, which is what the haymaker needs to know. Now I cut the hay when it is ready (when the clover is just beginning to bloom or the grass just beginning to head out) right after a rainy spell is over and hope the hay dries before the next round of wet weather. One moderate rain on cut hay doesn't hurt it too much, not nearly as much as if I put hay too green in the barn and spread it out in the loft to dry. In the latter case, the hay inevitably molds anyway and is no better than hay that gets rained on once in the field but then dries out before storing. If wet hay is piled up too deeply, it can heat and catch fire. Many barns have burned down this way. Some farmers sprinkle salt over hay that is not quite dry as they put it in the barn. I don't think much of this practice. I haven't tried citric acid.

After the hay is mowed, a decision must be made as to when it should be windrowed. The idea of the windrow is to lift the hay up from the swath so that it will dry faster. Long experience has proved to me that hay rarely dries faster in the windrow than in the swath unless the windrow is tedded, and I can't justify the cost of a tedder. Let the hay remain in the thin swath layer on the ground for the first half of the drying period. If rain falls, the hay will dry better in the swath than if rolled up in a windrow. Rake up the swath only for the last part of the drying process but before the hay is so dry that the raking shatters the precious leaves from the legume hay.

Haymakers argue endlessly over the fine points of good hay. Many believe that a first cutting, even if rained on once, still has more nutritional value than a lovely green third cutting from which half the leaves shatter off because the hay gets too dry before baling. In any event, alfalfa or clover hay, cut at the right time—that is, just as it starts to blossom—then dried quickly and gotten into the barn still somewhat soft and green, is a complete feed for all grazing animals and is almost always underpriced. Poor hay is almost always over-priced because it has very little value except as bedding. The most prevalent sin of husbandry is overwintering livestock on poor hay and trying to make up for it with grain. But sometimes one has no choice, because farmers have not known how or wanted to learn how to make hay in humid regions. Or the weather and their busy schedules prevent them from making good hay.

To learn how to judge good hay, go to a county fair and study the hay that wins blue ribbons. Look at it. Smell it. Feel it. Compare it to the samples that do not win. A farmer I know told me that once he made the perfect bale of hay. He took samples of it to the state fair for three years in a row and won a blue ribbon every time.

If you have dried herbs or flowers, you already know the perfect way to cure hay, even though using those methods on large volumes of forage is hardly practical. The herbalist knows exactly when a particular plant should be harvested both as regards the stage of growth and the time of day. The harvested herbs are then hung in a cool, dark (out of the sun) place, hopefully with air moisture not too high or too low. The colors of well-dried herbs look almost as bright as if they were still growing in the garden.

In humid climates, especially in Europe and in earlier times in America, very sophisticated, labor-intensive methods are used to get hay dried without rain damage. (In *Fertility Pastures and Cover Crops*, Newman Turner describes these methods in England in detail.) After the hay is cut, it is allowed to wilt for several hours, and then, using buckrakes on tractors (or the hydraulic front-end loader of a tractor, as I do), the hay is moved to and dropped on wooden or wire frames, which allow for ventilation under and through the little stacks that are thus formed. In some cases, a wooden tripod about eight feet tall acts as a stabilizer for the stack, and wooden trestles about two feet square are attached to the legs of the tripod in such a way as to form inverted, V-shaped little tunnels under the stacks. The hay is forked onto the tripod by hand from a pile of hay pushed up by the buckrake. The stack at completion is about ten feet tall and six feet through at the bottom, rising almost straight up and rounded over at the top so that water sheds readily. A field of these stacks is extremely attractive and, according to my view of life, that alone justifies the labor of making them. The smaller pasture farm might find them practical since livestock can self-feed from them in winter.

I confess to having fallen in love with haystacks. My grandfather made huge ones, straight-sided and symmetrical, a skill that I now understand is not easy to learn. Big stacks like his don't fit so small a pasture farm as mine, but little ones are extremely practical, by which I mean cheap and effective. I do not build mine over racks, as described above, since I can usually get hay stacked without too much rain. I am thinking, however, of devising a way to run perforated plastic pipes under and up through the center of my stacks for ventilation. I am satisfied that the pipe would work just about as well as the wooden racks.

I make stacks about the size of those described above, one in the center of each one-acre temporary pasture where I am making hay. I can carry the hay from the windrows close to the stack by hand fairly easily, and move only the outer three windrows to the stack with the front-end loader. To make a stack, I cut a slim pole from the woods about ten feet tall and set it in a hole dug with a posthole digger at the center of where the stack will stand. This pole acts as a stabilizer for the little stack. I need to put it in the ground only about two feet, because hay presses in from all sides to prevent it from leaning over. That's the ideal, of course, but sometimes I get a little careless and the pole (and the stack) lean a little. Unlike Monet, who had to worry only about color and light, I make haystacks with real hay, not paint, and so I have to pay attention to engineering principles.

After the post is in, I arrange a circle of woven wire fence around it about eight feet in diameter—I've never actually measured. The fence acts as a support for the first four feet of the stack, keeping it nice and straight all around. Then in goes the hay, usually by hand forking or from the front-end loader. After the hay rises above the ring of fence, I do the stacking by hand because I can't maintain nice straight sides on such a small stack with the loader. Windrows farther away from the stack we may load onto the pickup truck and drive to the stack. Standing on the truck makes building the stack easier.

To build a stack that is mostly vertical so that it sheds water well is an art. I don't claim to have mastered it yet, but I'm getting better. The secret is to pile the hay around the outside of the stack first.

Think of forkfuls of hay as you would pieces of firewood you are ranking up. Rank the hay around the outside of the stack, then throw forkfuls inside, next to the stabilizing pole, to pin in the vertical outside wall. By the time you get ten feet up, the stack will have edged inward, but only slightly. Then round off the top over the stabilizer with a big cap of hay. Legume hay is much easier to stack up than grass hay. Grass hay is slippery and wants to slide off the outer wall of the stack. Since almost all my hay is comprised of legumes, it stacks fine.

I make one stack in the shelter of the woods that is more than twice the size of the field stacks. The construction is the same, only larger, about sixteen feet in diameter and maybe fourteen feet tall. I would like to go higher, but I build this stack almost entirely with the front-end loader, and fourteen feet is as high as it will reach. The great blue heron that eats minnows out of our little farm pond nearby likes to perch on top of the stabilizing post sticking out of the center of this stack and keep an eye on me. But it is better to stack the last cap of hay over the post so rain can't run down it.

Okay, I have just told you the right way to make little stacks, the kind of thing that writers can do with words, just like painters can do with paint. The reality sometimes doesn't go so easily. In the summer of 2003, I built my little stacks as I had done other years. Then, instead of getting just a few showers over the summer, as is usually the case, we got five inches of rain in the first eight days of June. The rain did not shed off the stacks but beat them down. Before the stacks could dry out enough so that I could cover them with a tarp of some kind to save what remained of the half-ruined hay, five more inches thundered down three weeks later. The heavy rain smashed the stacks like a giant sledgehammer, and much of the hay rotted. If I had put a plastic cover over them, all would have been well, but I just had no idea what catastrophic rain can do to a little haystack. Now I do. Actually, the loss was not as bad as it sounds because the abundant rain produced enough additional hay in subsequent cuttings to make up for the ruined stacks. The hay in the stacks will now become compost to spread back on the fields or to use in the garden.

When I say that the animals self-feed the stacks, I am not being entirely accurate. They will eat awhile through the woven wire fence around the stack (you should use wire with eight- or twelve-inch stays so the sheep don't get their heads caught). Soon, however, they will make a mangled mess of the wire, so I take it off. In a few weeks, the younger sheep get a notion to climb the stack, just like children do. Older cattle and sheep eat great holes in the stack that the children use as caves. (My stack is not big enough to pose the danger of animals eating so far into it that it will collapse and suffocate them.) Sometimes all that is visible around our stack is a ring of sheep and cow butts. Left to their own devices, the animals would soon waste more hay than they eat. So I drive a few tall steel fence posts into the ground around the stack and hang wire cattle panels on them, raised up just enough so the animals can eat under the panels without climbing up the stack. Any number of wooden board designs could be devised to do the same thing. As the animals eat, the hay slides neatly and almost imperceptibly down the stabilizing pole. Children frolicking on the stack help. Sometimes too much. One of the most pleasant winter scenes on our farm is the congregation of animals and humans around the stack on a warmish afternoon. Chickens flutter all over the stack, finding various morsels to eat; children bounce on top of the stack; cows and sheep eat into the stack or lounge around it, chewing their cuds. Visualize an Edward Hicks barnyard painting.

Making Silage

Where early summer climate is so rainy that making that first cutting of hay on time is difficult, commercial dairymen resort to silage. Silage is green forage that is cut, wilted a little or a lot (a lot is better), chopped, and stored in a variety of ways to keep air out of it. Tall upright silos, once commonly seen on all farms, allowed the stored green grass to settle down by its weight so solidly that only a little spoilage occurred around the edges and on top. Then came sealed silos that kept the air out completely. Then came the much cheaper bunker and trench silos, where the silage was piled on the ground or in a trench and run over with tractors to pack it tight. Bunker and trench silos drew rats, and the livestock churned up mud in thaw weather. Concrete bunkers came into use, still not as satisfactory as upright silos but cheaper. A later storage method put the cut grass into huge plastic sacks that when filled look like enormous caterpillars lying on the ground. The silage was in this case not compressed because the plastic was supposed to keep out the air that would cause spoilage.

Now the most modern method is to make big round bales out of the cut and wilted grass and wrap each bale in plastic—all done mechanically. The forage is then referred to as balage and keeps very well, as long as the plastic wrapper is not punctured.

I have participated in all these methods personally, and I'd rather make hay. Balage really is making hay rather than silage, and I think it will become the prevalent way to provide forage other than pasture itself. It avoids most of the worry of bad weather ruining the hay. As long as winter or emergency feed is needed along with pastures, balage will be the preferred method of making it. Where hay is made to sell off the farm, standard baled hay will remain the choice because it is easier to transport.

When green silage ferments, it can give off a chlorinelike gas, especially in big enclosed upright silos. A whiff can kill you. I have elsewhere described my other reservations about silage. Mostly I think that hay or balage is better nutritionally for the animals.

I would not have considered balage a practical possibility for small grass farmers or garden farmers had I not come across a curious article written by an Austrian, Rudolph Seiden, in the Autumn 1950 issue of the now-defunct *Farm Quarterly*. He was advocating silage not only for very small-scale farmers but also for poultry producers! I will quote a paragraph from his article, "Pasture for Poultry" and ask you, before you read my reaction below it, if it inspires in you the same thought that came to me when I first read it:

> Poultry silage can be preserved from . . . young grass, or clover, to produce an easily digested protein feed that is rich in vitamin A and B. A simple method of preserving the grass is to cut it in one-fourth inch lengths; let it wilt to reduce the moisture content from around 80% to around 65%; then pack it into clean, air-tight barrels. Mixed with the chopped grass as it is put into each 50 gallon barrel are four gallons of molasses which have been diluted with 16 gallons of water. The molasses, rich in fermentable carbohydrates which green feeds lack, greatly improves the quality and palatability of the silage. The barrel lid, which must be somewhat smaller than the container's opening, is pressed down into the grass-filled barrel with 300 to 400 pound weights. As the grass settles, more of it must be added to keep the barrel full and tightly packed. After settling a few days, an airtight lid, or a double layer of tarpaper, is put on top of the container which should

then be stored for three months in a cellar or cool pit. When the silage barrel is opened, it will be found that the top layer will be slightly mouldy and should be discarded. The cider-like smelling ensilage that remains may be fed at the rate of two to five pounds daily to each 100 laying hens and up to 12 pounds daily per 100 turkeys.

I don't know about you, but my first thought was that Mr. Seiden was unwittingly suggesting a way to make something sane out of American lawn madness. We can turn our thirty million acres of lawns in this country into succulent meat, eggs, or milk. Instead of sending zillions of tons of lawn clippings to the landfill or the municipal compost yard, feed them to livestock. Farmers in or near every city could create a business out of it. The lawn mower chops the clippings into one-fourth-inch pieces. The clippings can be allowed to wilt a little on the yard and then be picked up by leaf and grass vacuum attachments on the lawn mower. In the plastic bags, the clippings could ferment in airtight safety. The suburban residents would bear all the cost of producing the feed and then pay the cattle feeder instead of the city trash collectors to take it off their hands. The lawn lover could buy the clippings back as meat at a discount for participating in the operation. It would be the first cattle-feeding venture in history that was profitable without a tax dodge or a subsidy.

My second thought was that making small amounts of silage is a lot like making sauerkraut, just as making good hay is a lot like drying herbs. The gardener already knows most of the skills and secrets involved.

For livestock, the molasses in Seiden's silage would work nicely, but in balage it's not necessary. A little grain would be better, I think, and certainly cheaper. If you don't have barrels or don't want to mess around with plastic bags, Seiden's article included a photo of a man forking grass clippings out of a wheelbarrow into a round tank about six feet in diameter and about four feet tall—a miniature upright silo. Another man was arranging the grass clippings evenly over the surface of the tank and tramping them solid. I imagine that both of them periodically got on top of the silage and bounced around to get a good tight pack. I imagine further that when they had filled the little tank they covered it with something to make it more or less airtight. Then, in winter they would fork the silage out into the wheelbarrow and feed it to their chickens. What a wonderful garden farm idea. With thirty million acres in lawns, and the acreage growing by leaps and bounds, this is the kind of thinking that must take place unless we want to buy all our meat, milk, and eggs from Brazil, where our grain monopoly is moving to.

21

Trees in the Pasture

Like nearly everything else, trees are a source of spirited debate among grass farmers. Those who are competing in the mainstream markets look to gain every advantage they can to make grass and legumes produce as much milk or meat in a given time as grain does. *Some* of these commercial producers look upon trees as a threat to production because they feel that livestock spend too much time on hot days standing in the shade instead of out in the pasture eating. These producers argue that heat doesn't hurt livestock, just as I argue that cold weather doesn't hurt livestock. They say this is especially true if you breed heat tolerance such as that exhibited by Brahmas into your animals. Also, animals huddled together under a tree or in the woods drop excessive amounts of manure there instead of scattering it out over the pasture.

I would not live up to my convictions or my reputation as a contrary farmer if I agreed with that point of view. (Also, I know a lot of graziers who agree with me.) I've not raised Brahmas, so I can't speak with authority on heat tolerance. But every cow, sheep, horse, pig, or chicken I've ever raised, of whatever breed or mixture thereof, sought the shade on hot summer days, just as I do. Some of them will seek shade on cool days if the sun is shining brightly—including Brahma-humped Brangus. And of their own accord animals will sleep out in the snow instead of going into the barn. I can't accept any human study that says animals don't know what's best for them in these situations.

The main reason dairy cows in the Deep South have a hard time competing with cows in Wisconsin in milk production is the heat. Southern dairymen have tried various ingenious forms of air-conditioning to solve the problem. I am sure no dairyman is going to let his prize milk cows stand out in the blazing sun for any length of time. I'm not accusing graziers of cruelty. Many times I have played in softball tournaments when the temperature was around one hundred degrees on the surface of the diamond. I was having a good time. But after a while, one just doesn't play very good ball in that kind of heat, and I doubt a cow or even a steer plays their best game in that situation, either, even if grazing on manna from heaven.

I have closely watched my animals on hot afternoons. There is tree shade handy along the borders of all my paddocks. The animals will graze awhile, move back into the shade for about half an hour, venture out and eat for another fifteen or twenty minutes, move back in the shade for another half hour, and so on. I am sure they are eating about as much as they do on a cool spring or fall day, and, if not, they are making up for it at night. Moreover, this is the time they clean up the weeds and brush that dare to venture into the shaded fencerows. I particularly notice how the animals like the fencerows lined with red cedar trees. They can walk under the low branches of these trees and brush off flies. I like to think that the cedar oil that gets rubbed onto their skin as they do so also discourages the flies a little.

Some graziers plant (or leave uncut) a scattering of shade trees over their pastures. As the earth moves around the sun, the shade from these trees moves also, and so the animals seeking that shade move around a little, too, rather than dump their manure in a few spots. The moving shade and the scattering of trees discourages crowding in any one place.

On the other hand, fence-line trees do not go well with electric fences. Limbs, if not whole trunks, fall on the fence. Low limbs grow over the fence. No good. Only because I have all woven wire fences and will replace them, if need be, with even sturdier wire panel fence, can I grow trees on my borders.

I can point out a disadvantage of cedar-lined fencerows that I doubt anyone would think of. Just this morning I witnessed another first on our farm. There were two young deer prancing around the near pasture seemingly unafraid of my approach. Believe it or not, they couldn't yet sail over the fences like the deer usually do, or they hadn't got the hang of it yet. Where they could see the fence, they would try to jump over but hit the top barb and fall back. So they turned and raced for the cedar-lined fence. They could see the trees, of course, but not the fence hidden by cedar branches. They hit the fence going full speed. It didn't seem to hurt them, but they broke off the fencepost. I opened a gate or two for them, but they never found the openings. Later they disappeared, so I presume they found a way out. If not, we may end up with two domesticated deer and a freezer full of venison.

One of the greatest potential advantages of pasture farming not yet much talked about is that it can allow the simultaneous pursuit of orcharding. A considerable number of fruit and nut trees can be grown on larger pastures without decreasing significantly the amount of pasture grass. In fact, the grazing animals come in handy for cleaning up dropped fruit, or they can be allowed to eat some or all of the fruit as part of their regular fare. Apple trees seem particularly suited for pasture orcharding. The animals love apples and keep them cleaned up as they fall. I have noticed that our pasture apple trees sometimes have blemish-free fruit without spraying, possibly because livestock eat the drops and the worms in them. We have learned to grow peach trees where the chickens range freely. The hens scratch around the trees and are apparently controlling the peach tree borer.

Remember that you can't plant trees in a pasture unless you protect them until they outgrow the reach of cows and horses. One of the reasons that I planted some fencerows to red cedars is that animals won't eat them. That can be risky. If animals won't eat the trees you plant, the tree seeds may eventually cause you a weed problem.

Here is a list of trees I like or dislike in pastures, purely from my own experience and biases. You will want to make your choices based on what grows best in your climate. As much as possible, especially on small pasture farms, be partial to species that can serve more than one purpose for you. A tree that just gives shade is not as beneficial as one that gives shade and food too. Or shade, food, and eventually firewood or good lumber.

I mentioned apples and peaches. All the temperate zone fruit trees should be considered. (Citrus fruits are not particularly liked by livestock, hogs, and chickens in my experience, but there may be some kinds, unfamiliar to me, that would be practical.) I would not buy expensive fruit trees for pasture orcharding unless I had in mind a particular variety I wanted to develop for the people market. Apple and peach trees can be grown easily from seed. The seedlings may not resemble the parent variety, but often these wild trees bear fruit just as good. You might even get fruit from a seedling with better quality than the parent. When I planted my pasture apple trees, I merely scattered buckets of cider pressings, full of apples seeds, along the fencerow in the late fall. So many grew that I had to thin them out. Or rather, I protected the ones I wanted to keep and let the livestock eat the others. Of course if you have animals grazing apples or any fruit, you may find a wild seedling come up wherever the livestock defecate.

I have not had much luck getting a quality cherry from seedling trees. Nor are the fruits of sufficient size to be particularly beneficial to the animals. The leaves might carry enough hydrocyanic acid to make them poisonous like wild cherry leaves when wilted.

Pears do not grow from seed as well as apples do for me, but the fruit makes food as good for livestock as apples. Pears, at least here, have the advantage of being less prone to insect attack than apples. A scattered grove of them in a pasture could be a practical orchard possibility.

In any event, where there is a choice of location, fruit trees ought to be planted on hillsides. The bigger the hill the better. The reason is that airflow down the hill on frosty nights in spring can keep the temperature just high enough to prevent frost injury to the blossoms. This is particularly true of peaches, which seem to have a death wish for blooming on the frostiest spring night of the season.

I have planted hickory nuts from selected trees in paddock corners where I can place a wire panel from one fence to the other out about six feet from the corner to handily prevent livestock from eating the young trees. These nuts come from outstanding trees. My friend in Iowa, Maury Telleen, once sent me some hickory nuts as big as eggs. With the husks still on, they were as big as baseballs. Large-sized hickory nuts often have very thick shells, but these did not. I had never seen such nuts. Fortunately, three of them sprouted in my fence corners and I hope to live long enough to see them fruit. Livestock won't eat hickory nuts, to my knowledge, but because of the size and ease of cracking, these nuts have market possibility. To me no nut tastes as good as hickory.

Farther south I would of course substitute pecan trees for hickory. Pasture farming and pecans have long been a traditional combination along the lower Mississippi River. Most of these farms have been plowed up for corn: an agriculture hell-bent on destroying itself.

Despite the fact that oaks are said to be toxic to cattle, another tree I like for pastures is chinquapin oak (*Quercus muehlenbergii*), not to be confused with chestnut oak or the smallish chinquapin of the chestnut family. Chinquapin oak is not a common tree though widespread throughout the Appalachian region. The acorns are not bitter like other acorns and make excellent mast for grazing livestock and hogs as well as wild animals, especially wild turkeys and deer. Our sheep clean up every one of these acorns from massive trees that in a good year produce bushels of them. Hogs relish the acorns of other oak varieties, too. J. Russell Smith in his classic book, *Tree Crops,* says that up to half of the feed ration for chickens can come from crushed or cracked acorns.

Black walnut trees grow very easily here from the nuts. Watching the nuts fall, embed themselves in the sod over winter, and sprout in the spring is a great way to convince yourself that winter-sown seed need not be planted in cultivated soil. (As I explained earlier, the notion that nut trees get started because squirrels "plant" acorns is a case of misinterpreting cause and effect.) Since black walnut wood is quite valuable, I at first planted some (dropped the nuts on the ground and stepped on them in December) in the fencerows and allowed many more to grow in the wild paddock (E) across the creek. Livestock will eat the leaves but not readily. Some herbals say that black walnut leaves are an effective wormer for livestock. But I no longer plant walnut trees in fencerows. Limbs regularly break off in windstorms and land on the fence. Also, livestock do not eat the nuts. We gather a bushel of them for our own table use.

Black walnut trees work well out in the pasture because the leaves are fine and do not cast a heavy shade. Also, the tree roots exude a substance that discourages weed growth around the trees, but strangely seems to encourage bluegrass. Commercial black walnut growers have learned to combine pasture farming with walnut tree farming because of these advantages. Obviously, other walnuts of commercial food value could be substituted for black walnut.

Many trees share with black walnut a propensity to drop limbs on fences. Weeping willow is one of the worst. I planted two next to each other in a fencerow because, well, for reasons of the soul: because they green up early in spring and stay green until late in the fall. They are also easy to start. Just stick a cutting in the ground in the spring and it will usually root. But I regretted the decision and my neighbor regrets it even more. For twenty years, no problem. Then the trees started losing limbs—big limbs. And they usually managed to fall on the fence—the line fence in this case. The look my neighbor gives me says what he is thinking: now you know why I don't like trees in fencerows.

I have long held a vision of a hedgerow fence that would hold in livestock like the ones in Europe. Those hedgerows were the work of centuries and not easily duplicated in a lifetime. They not only stopped cattle for centuries but also tanks in World War II. That is a definition of a really good fence: one that stops war tanks, literally and figuratively. The only person I know who is trying to make an American version of European hedgerow fence is my friend Brian Knopp, a shepherd in North Carolina. He uses a wild orange that is related to Osage orange (once a chosen tree for hedgerows because of its vicious thorns) and barberry. He doesn't guarantee his fence but has had some success with it.

After many unpleasant adventures with multiflora rose, the so-called living fence, I tried honey locust, which has also been proclaimed for hedgerows. But honey locust wants to grow rapidly upward, leaving open space below for animals to walk right through. To keep it growing low and bushy requires much yearly trimming. I suppose after a century one might have a good fence, but in ten years I had nothing to show other than a whole bunch of thorns that have a propensity for finding my tractor tires.

On the other hand, thornless honey locust makes an excellent pasture tree, even if it is not good for living fences. It produces the same sweet seedpods that regular honey locust produces. Livestock love these pods when they fall in late autumn and can graze under the thornless trees without the danger of stepping on a thorn. Moreover, the tree is a legume, so it is constantly fixing nitrogen in the soil. Most of all, the leaves of the tree are very fine and do not cast heavy shade. Thus the grass under them grows vigorously not only because of the added nitrogen but also because the light shade both allows sufficient sunlight and keeps the soil from drying out as fast as in the open sunlight. *Tree Crops*, cited earlier, devotes many pages to lauding the potential value of honey locust beans for animal feed. I have eaten both the sweet flesh inside the pod and the green seeds cooked like peas. Some books say the flesh is toxic, but I am still going strong. The cooked seeds are okay, but peas are a whole world better.

Black locust, although possessing some of the advantages of honey locust and in addition having wood that is almost impervious to rot, is not a good pasture tree. Its leaves are considered toxic to animals.

I once made the mistake of planting a row of wild hawthorns (white thorn to farmers hereabouts) along the side of our property. I figured that if I planted the thorny devils close enough together, I could entwine their branches and make a fence that would stop a Humvee, if not a tank, and certainly cattle. True enough, I suppose, but wild hawthorn quickly becomes a terrible weed. Livestock and wild animals eat the little red fruit and everywhere their droppings fall, hawthorns spring up. You can mow hawthorns off for fifty years, as my cousin has, and they keep right on growing back. I realized my mistake soon enough, especially after my cousin, who is the most kindly man I know, scolded me roundly. It took a few years to get rid of them (with chemical brushkillers) and thereafter constant vigilance. Too bad, too, because they are beautiful in bloom or in fruit.

I decided to plant red cedars (common juniper) along the borders of some of the paddocks, as I have said. I was by then beyond all romantic notions of developing a living fence, but if the cedars would not make an impenetrable hedge, they would provide a source of fence posts, a most important advantage for a pasture farmer. Also, red cedars are especially good habitat for many birds and provide shade and windbreak for livestock. Plus, I could plant a red cedar hedge without devising a way to protect it from grazing animals while it was young. Livestock won't eat the trees unless they are starving.

I have been pleased with my red cedars. They tend to stool out into several trunks, which makes for a thicker hedge. The stool trunks occasionally break off in wind or especially in ice and snow storms, but they are so bushy that they seldom harm the woven wire fence even if they sag right on it. These stool trunks are usually big enough to use as fence posts, too, so there is little waste. Branches break off from snow and ice, too, but these are of manageable size and are fairly easy to clear out and burn. After twenty-five years, the trees, planted about twenty feet apart, have closed in the space between them, providing an excellent windbreak.

Red cedar wood has many uses. In addition to making enduring fence posts, the wood is also in demand from woodcarvers and woodworkers. One of my brothers-in-law makes beautiful boxes of red cedar wood. It is, of course, even more in demand for mothproof closets and chests. If you have priced a cedar-lined closet lately, you know you are dealing with very high-value wood. Moreover, the trees are comely, green in winter and summer. If you travel in Kentucky, you will find many horse farm pastures dotted with red cedar trees, a very pleasing landscape.

To keep a hedge of red cedar looking nice requires yearly maintenance to cut out grapevines and poison ivy and other kinds of seedling trees that try to gain a foothold under or next to the cedars. On the side of the fence where the sheep seek shade, they can clear out most of this brush, but on a line fence you will have to patrol the other side and keep things in order.

There is a downside to everything. Red cedars produce zillions of berries, which birds love (but which unfortunately aren't big enough to be practical for making gin). We think the overwintering bluebirds we now enjoy don't go south anymore because of the ample supply of cedar berries. The birds eat and scatter the seed in their droppings. New trees come up everywhere. They can become a weedy pest.

But control is relatively easy (compared to multiflora rose, for instance). If the seedlings are mowed off when they have grown a foot tall, they will seldom grow back. Regular lawn herbicides or Roundup control them, too.

Livestock will graze the freshly fallen leaves of sugar maple and other trees. Red maple leaves, however, are supposed to be poisonous to livestock. Here again I am at a loss. We have allowed sheep and cows access to our woodlot in the fall for years to eat both fallen leaves and acorns. Many red maples grow in this woodlot and we have not had problems with toxicity. Sugar maple makes sugar, of course, and the wood is currently becoming more valuable as furniture wood. But maples throw such heavy shade that grass won't grow under them, so it is best to grow these trees in a woodlot, not pasture. The animals like fresh green maple leaves, too, and stretch their necks upward to eat all the maple foliage they can reach. I have seen sheep straddle maple seedlings and ride them down to eat off the leaves. Since myriad maple seedlings come up every spring in our woods, a sufficient number of saplings grow above grazing height for future trees. Red maple seedlings also grow in the woods, although not as profusely as sugar maple. Again, all I can say is that we haven't experienced problems with toxicity.

Every pasture farm needs a woodlot, not for grazing but for windbreak or winter shelter. If you have ever walked from a windy winter pasture into a thick grove of evergreens, you know that even a calving cow needs no more shelter than that.

22

A Walk on the Wild Side

When a pasture farmer makes a paradise for farm animals, he also makes a paradise for wildlife. But appreciating the overall ecological benefits that wildlife brings to the farm is not always easy. As I mutter dire threats under my breath at the sight of a lamb killed by a coyote, I have to keep reminding myself that the coyote is also decimating the population of groundhogs that in our farm's years BC (before coyotes) had few effective predators. While I rail against the burgeoning population of crop-destroying deer, wild turkeys, Canada geese, squirrels, and rabbits, I am also sure that we could get all of our meat from these creatures if we wanted to, which suggests that I might be insane for bothering with domestic animals at all.

Being able to observe the wonders of wildlife in and over our pastures is a treat well worth paying for with a few lost lambs or chickens. The enjoyment begins with creatures whose good work is hardly recognized by the human race in general but held in reverence by the pasture farmer. Walking in April twilight where the soil surface was still partially bare because the clover and grass were only beginning to grow, I was suddenly aware of night crawlers slithering back into their holes at my approach. I stopped and looked ahead of me. As my eyes grew accustomed to seeing the worms in the waning light, I realized the ground was literally covered with them, so many, in fact, that my first reaction was one of repugnance. It looked as if someone had tried to cover the soil surface with spaghetti. From reading, I knew that earthworms could produce up to sixty *tons* of castings per acre per year in optimum situations. Pasture farming on fertile soil is as optimum a situation as can be attained. Surely I was looking at a sixty-ton-per-acre potential. These castings can't add much more in soil nutrients than the steady increase of humus and minerals wrought by sun and rotting organic matter, but the castings represent a refinement of those nutrients, making them much more available to plants than they otherwise would be. There is in addition the nutritional value of the earthworms themselves when they die, which, in a population such as I was looking at, translates into a couple more tons of nutrients per acre. Also, the burrowing worms increase soil aeration, allowing more rain to soak in instead of running off. I was looking at another example of what Charles Darwin meant when he said that the earthworm is one of the major saviors of life on earth. And I didn't have to raise a finger or spend a dollar to produce all this soil enrichment.

When it came time to rotate that paddock to corn in my scheme of things, I wondered if perhaps I should use some added fertilizer. I had not applied any except manure for a decade, and I had harvested hay from the field for the last two years, a practice that supposedly depletes the potassium content of the soil. Remembering all those earthworms and remembering Darwin's statistics, I decided to add nothing to the soil. To make matters seemingly more risky, rain prohibited me from getting the corn planted until June. Would it grow properly?

Now, in August as I write this, that corn is "as tall as an elephant's eye," as the song goes, and has the deep green color that corn achieves only when it is growing on soil of the highest fertility. If I had added fertilizer, I would have to harvest the ears with a cherry picker.

Obviously, a boy of the hayfields and pasturelands could be harvesting some of those night crawlers and selling them for fish bait. In the quantities that occur on established pastureland, he could earn some pocket money.

The second most useful and beneficial form of wildlife on our farm is the pollinating bee, especially honeybees and bumblebees. Bumblebees are the best pollinator of red clover that we have, and it is important for us to get as much red clover seed as possible to reseed the stands.

The honeybees' domain is white clover, our most important pasture plant in turning grass to flesh. Some of these bees come from the hive I keep; some, from hollow trees. I love to sit in the pasture and watch them work for me. While they gather honey for us, they pollinate the blossoms, making sure that enough seed is produced to keep the clover reseeding itself year after year with no help from me. The clover is putting as much as two hundred pounds of nitrogen per acre to the soil, again without my lifting a finger or opening a billfold. I sit there, bees around me, earthworms below me, and I wonder why humans can't seem to see how easy life could be if only we would let it be.

When making hay, I always find garter snakes under the windrows. There is hardly a creature more beneficial in terms of eating farm and garden pests. Also in residence in the pastures are leopard frogs and green frogs, which get it into their heads in summer to leave the safety of creek and pond bank, I guess to gorge on grasshoppers and crickets in the meadows. Most bizarre of all, crayfish make their way from the creek into the adjacent field and actually graze on clovers and grasses and especially oats if any is available. Being more at home in water, they burrow down in the soil to reach it because the water table in this pasture often rises to within two feet or even less of the surface. They build telltale little mud chimneys above their holes, and if I were hungry enough or energetic enough, I could dig up a batch of these miniature lobsters for a gourmet meal. In the South, where crayfish are larger, these mud chimneys can be a nuisance when mowing hay.

The air above the clover is alive with more than bees. A flash of azure explodes from a hole in a fence post: a bluebird. His royal highness, the kingbird, sits on the fence in that lordly manner for which he gets his name. In the interest of political correctness, should we call the females "queenbird"? A red-winged blackbird hovers above the red clover and then settles into a nest I cannot see. I make a mental note of the location so that I can watch out for it when I cut hay. I sometimes cuss the cowbirds, which lay their eggs in the nests of other birds, allowing the fledglings to literally eat the resident birds' young out of house and home. But today I watch them approvingly as they flock around the sheep, eating horseflies and other livestock pests.

Hawks float across the sky, looking for mouse movement in the grass. Buzzards wheel above me, my immobility suggesting that I might become their next meal. A song sparrow sings, that most melodious of all birdsong. The two other champion songsters, the meadowlark and the bobolink, left us a few years ago as pastureland in corn country dwindled. If grass farming continues to spread, they will be back. They will follow the grass back.

In my view, the most unsung of the pasture birds is the buzzard. It doesn't sing at all. Buzzards clean up dead animals and in doing so save the pasture farmer considerable time and money over the years. Even in a smaller flock of sheep, there is always one of them on the verge of death. The buzzards make short work of carrion. In the process, they are awesome to watch. Several times I have seen as many as twenty of them on consecutive fence posts near where they were demolishing a dead animal. At my approach they spread their wings to their full six feet span and then raise them to a horizontal position. It is an awesome sight. Whether they are trying to scare me off or just getting ready to fly should I come closer is hard to say. But they freeze in that position, and as long as I do not advance, they make no move to fly. Twenty great black, bald-headed birds, wings extended six feet, perched on fence posts like so many Kwakiutl Indian thunderbirds atop their totem poles.

Vying for my attention in the summer pasture are butterflies: the monarch, the black swallowtail, the tiger swallowtail, and a score of smaller but no less beautiful species. Some of them are drinking clover nectar and paying for their treat by pollinating the clover, too.

For me, the most gratifying insect of the grassland is the firefly, winking, blinking in the dusk. Meadows are the favored habitat of fireflies, especially the Pyralis. Nor is their beauty their only gift to us. Their larvae eat the larvae of other insects, especially slugs and mites. A few years ago, an Amish family I know found a market for fireflies from a company investigating the properties of the chemical that fires their light. For several years, the family made extra income supplying that market.

The insect that should rule the pastureland, the dung beetle, vanished from our farm and the farms around us about the same time that pastures started to disappear. When I was a child, dung beetles were a favorite form of entertainment. We would mark a line across a sheep path and bet which pair of toiling beetles would reach it first as they pushed their little marble-sized manure balls along. In June, the grass was literally alive with them, eating cow and sheep manure, forming their manure balls, inside of which the female laid an egg, and then pushing the balls to another location for burial. (Some dung beetles dig burrows first and then carry blobs of dung down into the burrow, where the female forms them into balls and lays an egg in each.) The eggs hatch, the larvae eat their way out of the manure ball, pupate, and eventually tunnel to the surface as adult beetles and repeat the life cycle. The pasture manure mostly gets buried in the process. In those days, there was no worry about manure not deteriorating fast enough on the sod surface. The dung beetles took care of that.

Where did they go? I can't get a satisfactory answer from anyone. My own guess is that they declined as animals went off pastures into confinement. Their death knell sounded with the adoption of systemic livestock wormers that killed more than just internal parasites in the manure. Ironically, the wormers did not kill off the parasites.

But in many parts of the country, dung beetles still flourish, thanks to ranchers like Walt Davis, famous for his successful efforts at ecologically sane cattle ranching in Oklahoma. Information is available from both the University of Texas, Austin (Pat Richardson, Department of Zoology, was in charge last time I wrote about dung beetles), and Texas A & M, College Station (Truman Fincher). These people have mounted ongoing efforts to reintroduce dung beetles to cattle ranges. Ms. Richardson, on the phone a few years ago, described to me how Mr. Davis, after experimenting with some eleven different kinds of dung beetle, concluded that *Onthophagus gazella* was the most effective species for his purposes. They kept his three thousand acres of pasture clear of manure from early June through September. Behrens figured that the beetles were burying two thousand pounds of wet manure per acre per day. The dung beetles not only turned the manure into soil nutrients but also got rid of the breeding habitat of horn flies and face flies.

On a pasture walk one evening, I watched a skunk sauntering across a paddock, flipping over dried cow pies in search of earthworms. Raccoons do this, too, and, I imagine, opossums. There is, in fact, a whole little world of nature radiating out from cow pads. The woodcock has learned how to overturn cow pies and poke its long beak into the soil under them to snag night crawlers. Once in early spring, with the snow still melting, I watched bluebirds sitting on a fence post, darting down to the ground occasionally. I investigated. In flashes of sapphire, the birds were landing on old cow pies—miniature islands surrounded by snow melt. The sun had warmed the cow pies enough so that some little gnatlike bugs were emerging and hovering above them. The bluebirds were nabbing the bugs.

Not much is made yet of the treasure trove of meadow mushrooms that follows the grazing of livestock, especially horses and sheep. After a few years of manure rotting away in pastures, an abundance of the white meadow mushroom (*Agaricus campestris,* often called the Pink Bottom) began appearing in the fall, especially where the sheep had congregated. This mushroom is closely related to the common commercial mushroom, and, gathered fresh when the gills are still pink, it is absolutely delicious if you enjoy mushrooms. Sometimes, but very rarely, a poisonous amanita mushroom that looks like the Pink Bottom will grow in pastures, but its gills are pure white. The Pink Bottom often grows in our pastures in charming "fairy circles." The circles are never perfect, but quite distinct. I am surprised our dipsy society hasn't decided that aliens from outer space are at work here. There are other *Agaricus* species that grow in pastures, especially the *A. arvensis,* or horse mushroom, which is also delicious.

Delectable puffballs often grow in pastures, too, some as big as basketballs. Enterprising farmers do sell these in some farmers' markets, especially in Michigan. It does not take many basketball-sized puffballs to have enough to sell. A couple of slices, fried, make enough for a meal.

Obviously, in dealing with fungi, one must be careful. There is one puffball that is purple on the inside. Common sense would dictate to stay away from it. If you don't know which mushrooms are good and which are not, make friends with a mushroom hunter. The fact that he is still alive is good evidence that he knows what he is doing. An old puffball, as all country children know, makes an explosion of dusty powder when tramped on vigorously. I save a little of that powder on occasion for use as a healing agent. If you have to de-horn cattle, puffball powder and spider webs make a poultice to stop the bleeding. Yeah, yeah, I didn't believe it either until an old farmer in Minnesota, Ed Hesse, proved its effectiveness to me.

A pasture can produce garden vegetables, too. Just recently, Debbie Apple of the no-grain dairy featured in chapter 3 was telling me that a relatively new organization, the Slow Food Association, had gotten in touch with her. They were interested in more than no-grain milk. Some members liked lambsquarters for salads well enough to pay to get some. As I have said, it is hard to keep lambsquarters growing in a permanent pasture because livestock like the weed so much, but it will keep occurring there and in pastures being reserved for hay. The same for redroot or amaranth, another wild vegetable treasured by gourmands.

Thinking of a pasture as a garden extends also to flower gardening. In early spring our permanent pastures are literally covered with spring beauties. They are followed by my favorite pasture flower, blue-eyed grass. Both persist as long as pastures are rotated so that grazing is not continuous from April to July. Where paddocks are rotated to corn and hay occasionally, all kinds of wildflowers that are too pretty to call weeds will grow. The evening primrose tops the list, but the purple blossoms of ironweed and goldenrod almost rival it. Beating all three are spiderwort, the purple New England aster, and bergamot.

Humans have really only begun to investigate the possibilities that a pasture farm might turn into profitable products. Recently, scientists learned that mayapple contains an anticancer drug like its South Asian cousin, which has been harvested nearly to extinction for that reason. Mayapple grows in open wooded pastures. Grazing animals will not touch it. Scientists have found a way to extract the drug cheaply from the plant. Floyd Horn of the Agriculture Research Service says that the plant "could turn out to be a new alternative crop for U.S. growers." The fact that livestock won't eat it means that it can be grown along woodlot borders in concert with grazed pastures.

To name all the myriad lives, botanical and biological, that find home in the meadow would bore the reader, I fear. And most of these lives I do not even know yet. I walk my pastures enveloped by them all, finding on every walk something new or something reassuringly old. I sit at the top point of the pasture hill, look over my little domain, and wonder why I have been so blessed to be here and blessed even more by knowing for certain that I do not want to be anywhere else. I think of my favorite lines of poetry, from Wendell Berry's poem "On the Hill Late at Night."

> . . . I am wholly willing to be here
> between the bright silent thousands of stars
> and the life of the grass pouring out of the ground.
> The hill has grown to me like a foot.
> Until I lift the earth, I cannot move.

Annotated Bibliography

Periodicals

Farming, a quarterly magazine devoted to family farming, with a strong emphasis on grass farming. Box 85, Mt. Hope, OH 44660, http://www.farmingmagazine.net. This is my favorite farm magazine because, well, because I write for it.

Graze, a monthly magazine on grazing, P.O. Box 48, Belleville, WI 53508, http://www.grazeonline.com/.

Stockman Grass Farmer, a monthly magazine edited by Allan Nation, P.O. Box 9607, Jackson, MS 39286, http://www.stockmangrassfarmer.com/ sgf/. The best all-around source of information specifically on grazing.

All farm magazines now carry occasional articles on grass farming, rotational grazing, and year-round grazing. In addition, almost all the popular consumer magazines are publishing articles on meat, milk, and eggs from grass-fed animals. One I found particularly interesting is "Splendor in the Grass," by Sam Gugino, in the August 31, 2003, issue of *Wine Spectator,* because it gives some detailed assessments from high-end restaurant chefs about the taste of grass-fed beef versus corn-fed beef. It also includes (as most magazine articles do) sources of grain-fed meat, such as Conservation Beef, Helena, MT, 877-749-7177; D'Artagnan, Newark, NJ, 800-327-8246; www.eatwild.com (lists sources nationwide); Napa Free-Range Beef, St. Helena, CA, 866-661-9111; New England Livestock Alliance, Hardwick, MA, 413-477-6200; and Sunnyside Farms, Washington, VA 540-675-3636.

Books

Hughes, H. D., Maurice E. Heath, and Darrell S. Metcalfe, eds. *Forages: The Science of Grassland Agriculture.* 2d ed. Ames: Iowa State University Press, 1962. Out of print. I got a copy from Powell's Used Books on the Internet. The book is now in its sixth edition, as *Forages: An Introduction to Grassland Agriculture,* under the editorial authorship of Robert F. Barnes et al., from Blackwell Press.

Lee, Andy, and Patricia Foreman. *Day Range Poultry: Every Chicken Owner's Guide to Grazing Gardens and Improving Pastures.* Buena Vistsa, VA: Good Earth Publications, 2002. A very practical book, easy and entertaining to read.

Berry, Wendell. "On the Hill Late at Night." From *Farming: A Hand Book.* New York: Harcourt Brace Jovanovich, 1971.

Moncrieff, Elspeth, with Stephen and Iona Joseph. *Farm Animal Portraits in Britain, 1780–1900.* Wappinger Falls, NY: Antique Collectors' Club, 1996. For ordering information, call 800-252-5231.

Murphy, Bill. *Greener Pastures on Your Side of the Fence.* Colchester, VT: Arriba Publishing. A good interpretation of the Voisin method for American farmers.

Robinson, Jo. *Pasture Perfect.* Vashon, WA: Vashon Island Press, 2004. Sequel to *Why Grassfed Is Best!* (now out of print). The hottest book right now on the nutritional value of grass-fed animal food products. Aimed at the consumer. Not much how-to information on grazing itself.

Salatin, Joel. *Family Friendly Farming: A Multi-Generational Home-Based Business Testament.* South Burlington, VT: Chelsea Green Publishing, 2001. A good source for all of Salatin's books and other pasture publications is Back Forty Books, Hartshorn, MO, 65479, 866-596-9982.

Thompson, W. R. *The Pasture Book.* College Station, MS: W. R. Thompson. Ten editions printed between 1949 and 1951. Out of print. Practical details on pasture farming not found in other books. Somewhat dated and written primarily for southern grass farmers. I love this book because of choice tidbits about farming that seldom get printed.

Turner, Newman. *Fertility Pastures and Cover Crops.* 2d ed. Pauma Valley, CA: Bargyla and Gylver Rateaver, 1974. Available from Acres U.S.A., P.O. Box 91299, Austin, TX 78709; 800-355-5313; www.acresusa.com. Another of my favorites because of details about farm practices I have not found anywhere else.

Turney, Henry W. *Texas Range and Pastures—The Natural Way.* Privately published. Available from Acres U.S.A., P.O. Box 91299, Austin, TX 78709; 800-355-5313; www.acresusa.com.

Voisin, André. *Grass Productivity.* 1959. Translated from the French by Catherine T. M. Herriot. Covelo, CA: Island Press, 1988. The classic book on rotational grazing.

Index